Military Mission Formations and Hybrid Wars

This volume explores and develops new social-scientific tools for the analysis and understanding of contemporary military missions in theatre.

Despite the advent of new types of armed conflict, the social-scientific study of militaries, in action, continues to focus on tools developed in the heyday of conventional wars. These tools focus on such classic issues as cohesion and leadership, communication and unit dynamics or discipline and motivation. Although these issues continue to be important, most studies focus on organic units (up to and including brigades). By contrast, this volume suggests the utility of concepts related to mission formations – as opposed to "units" or "components" – to better capture the (ongoing) processual nature of the amalgamations and combinations that military involvement in conflicts necessitates. The study of these formations by the social sciences – sociology, social psychology, anthropology, political science and organization science – requires the introduction of new analytical tools to the study of militaries in theater. As such, this volume utilizes new approaches to social life, organizational dynamics and armed violence to understand the place of the armed forces in contemporary conflicts and the new tasks they are assigned.

This book will be of much interest to students of military studies, sociology, security studies and international relations in general.

Thomas Vladimir Brønd is an Assistant Professor at the Royal Danish Defence College.

Uzi Ben-Shalom is Chair and Associate Professor of the Department of Sociology and Anthropology at Ariel University, Israel.

Eyal Ben-Ari is a Research Fellow at the Kinneret Center for Society, Security and Peace, Israel.

Cass Military Studies

Commercial Insurgencies in the Networked Era
The Revolutionary Armed Forces of Colombia
Oscar Palma

The Politics of Military Families
State, Work Organizations, and the Rise of the Negotiation Household
Edited by René Moelker, Manon Andres, and Nina Rones

Organisational Learning and the Modern Army
A New Model for Lessons-Learned Processes
Tom Dyson

Civil–Military Relations in International Interventions
A New Analytical Framework
Karsten Friis

Defence Diplomacy
Strategic Engagement and Interstate Conflict
Daniel H. Katz

Management and Military Studies
Classical and Current Foundations
Joseph Soeters

Understanding Insurgent Resilience
Organizational Structures and the Implications for Counterinsurgency
  Andrew D. Henshaw

Military Strategy of Middle Powers
Competing for Security, Influence and Status in the 21st Century
Håkan Edström & Jacob Westberg

Military Mission Formations and Hybrid Wars
New Sociological Perspectives
Edited by Thomas Vladimir Brønd, Uzi Ben-Shalom and Eyal Ben-Ari

For more information about this series, please visit: https://www.routledge.com/
Cass-Military-Studies/book-series/CMS

Military Mission Formations and Hybrid Wars

New Sociological Perspectives

Edited by
Thomas Vladimir Brønd,
Uzi Ben-Shalom and
Eyal Ben-Ari

Routledge
Taylor & Francis Group

LONDON AND NEW YORK

First published 2021
by Routledge
2 Park Square, Milton Park, Abingdon, Oxon OX14 4RN

and by Routledge
605 Third Avenue, New York, NY 10017

First issued in paperback 2022

Routledge is an imprint of the Taylor & Francis Group, an informa business

Publisher's Note
The publisher has gone to great lengths to ensure the quality of this reprint but points out that some imperfections in the original copies may be apparent.

British Library Cataloguing-in-Publication Data
A catalogue record for this book is available from the British Library

Library of Congress Cataloging-in-Publication Data
A catalog record has been requested for this book

ISBN 13: 978-0-367-56721-7 (pbk)
ISBN 13: 978-0-367-42715-3 (hbk)
ISBN 13: 978-0-367-85539-0 (ebk)

DOI: 10.4324/9780367855390

Typeset in Times New Roman
by Deanta Global Publishing Services, Chennai, India

Contents

Lists of figures   vii
List of tables   viii
Contributors   ix

PART I
Introduction and Reflections on the Field   1

1   Introduction: Mission formations and a new agenda for the study
    of military units in action   3
    EYAL BEN-ARI, UZI BEN-SHALOM, THOMAS BROND AND CARMIT PADAN

2   New directions in military sociology: Reflections on a book project
    15 years later   23
    ERIC OUELLET

PART II
New Organizational Forms and Processes   43

3   Organizational adaptations in the hunt for Abu Musab al-Zarqawi:
    A review of concepts for analyzing their usefulness   45
    WILBUR J. SCOTT

4   Bureaucracies, Networks and Warfare in a Fluid Operating Environment 64
    JESSICA GLICKEN TURNLEY

5   From leading combat units to leading combat formations:
    Modularity, loose systems and temporariness   91
    EYAL BEN-ARI

PART III
Methodologies for the Study of Military Formations   105

6   Research approaches to the study of combat formations: A
    Personal Note   107
    UZI BEN-SHALOM

PART IV
Glocalized Mission Formations   121

7   Institutional isomorphic change in South Korea's UNPKO mission
    formation   123
    INSOO KIM AND YOUNG-IL CHOI

8   "Democracy… 120 mm at a time": Mission formations and
    operational entrapments in post-9/11 Afghanistan   141
    THOMAS RANDRUP PEDERSEN

9   Logics battlefield: IT contracting and military reserves in
    the Dutch army   167
    JOSEPH SOETERS, GEROLD DE GOOIJER, PAUL C. VAN FENEMA
    AND NUNO OLIVEIRA

PART V
Bringing it all Together   181

10 Integrative epilogue: What's new about the mission formation
    approach? Thinking through the military–academic juncture 183
    THOMAS CROSBIE

    Index   191

Figures

6.1 A sketch of a mission formation during a squad-level skirmish   115
6.2 A sketch of an air-ground mission formation   116
6.3 A sketch of a mission formation during the disengagement from
     Gaza in 2005, organized top–down   117
7.1 The network of overseas military units with similar unit tasks   130
7.2 Distribution of peace operation units by political and economic
     condition   131
7.3 Organizational structure of the Lebanon Peacekeeping Group
     (June 2008)   136
7.4 Cumulative number of peace operation participants and academic
     journal papers about peace operation   138
9.1 Reserves and triadic relationships   168

Tables

7.1   South Korea's peace operation units (1991–present)   127
7.2   Composition of South Korea's UNPKO units   128
7.3   Distribution of South Korea's peace operation by year and country   132
7.4   Mimetic isomorphism between South Korea's overseas military units   135
7.A.1 Unit task of South Korea's overseas military unit (1991–present)   139
10.1   Scholarly Lessons Learned as Equivalent Military Policy Concepts   189

Contributors

Eyal Ben-Ari is Research Fellow at the Kinneret Center for Society, Security and Peace and has carried out research in Israel, Japan, Singapore and Hong Kong. He studies the armed forces (including gender issues and combat units), early childhood education, and popular culture in Asia.

Uzi Ben-Shalom is the Chair and Associate Professor of the Department of Sociology and Anthropology in Ariel University. He is interested in sense-making and behavior in combat and military leadership. He is a joint chair of the military and security community of the Israeli Sociological Association and the 10th working group of the European Research Group of Military and Society (ERGOMAS).

Thomas Vladimir Brønd is an Assistant Professor at the Royal Danish Defence College. As a specialist in political anthropology, military anthropology and Middle East studies, he has done ethnographic fieldwork among revolutionaries in Syria, Lebanon and France and among the Danish Armed Forces.

Thomas Crosbie is an Associate Professor of Military Operations at the Royal Danish Defence College. A sociologist by training, his work focuses on the political agency of senior officers, the history of NATO's military policies, the privatization of security and the education of military elites.

Gerold de Gooijer served for more than 30 years with the Netherlands Defense Organization (NDO) and participated in various operations in Europe and the Middle East before joining the Netherlands Defense Academy as a lecturer and professional PhD candidate. He has acquired comprehensive experience in and knowledge of (military) logistics. He is conducting PhD research on interorganizational relationships in the field of "Total Force", in order to better understand internal NDO adaptation, changes to bilateral cooperation between NDO and other (types of) organizations, and logistics ecosystems.

Insoo Kim is a Lieutenant Colonel in the South Korean Army and a sociology professor at the Department of Political Science, Korea Military. He earned a PhD in sociology from the University of Wisconsin at Madison. His research interests cover democratization, military organization and military culture.

Nuno Oliveira (PhD, LSE) is an assistant professor in Organization Studies at Tilburg University, the Netherlands. His research focuses on core problems of organizing between organizations and has appeared in *Academy of Management Annals*, *Journal of Management*, *Journal of Supply Chain Management* and *Organization Science.*

Eric Ouellet has a doctorate in sociology from York University, Toronto, and is a full professor at the Royal Military College of Canada in the Department of Defence Studies. His research focuses on institutional theory and organizational studies applied to non-conventional aspects of warfare, and military affairs in general.

Carmit Padan is a senior research fellow at the Samuel Neaman Institute of Technion and a research fellow at the Institute for National Security Studies of Tel Aviv University. His main areas of research are military leadership, civil–military relations and national resilience. His work on these subjects has been published in journals and edited volumes and has been presented at international academic conferences.

Thomas Randrup Pedersen is an assistant professor at the Royal Danish Defence College. He received his PhD from the Department of Anthropology at the University of Copenhagen in 2017. He has carried out "embedded research" with Danish troops in Afghanistan, Iraq and Denmark.

Wilbur J. Scott is Professor Emeritus, Department of Sociology, University of Oklahoma, and Department of Behavioral Sciences & Leadership, United States Air Force Academy. His central areas of research are the sociology of violence and war and the sociology of veterans' issues.

Joseph Soeters is a professor of organizational sociology at Tilburg University. He was previously affiliated with the Netherlands Defence Academy. His most recent book (2020) is about the connection between management thinking and the study of the military, both in peacetime and in operational conditions.

Jessica Glicken Turnley (PhD, Cornell 1983) is a cultural anthropologist who has worked with many players in the US national security complex, including the nuclear weapons laboratories and military services. She also has worked with organizations in the private and non-profit sectors.

Paul C. van Fenema (PhD, Rotterdam School of Management, Erasmus University) is a professor of military logistics and organization science at the Netherlands Defense Academy. His interests include process research, management concept development, digital transformation/intelligent logistics, value propositions/business models and interorganizational innovation. Paul has published in journals such as *MIS Quarterly*, *Organization Science*, *Communications of the ACM*, *International Journal of Management Reviews* and *Industrial Marketing Management.*

Young-Il Choi has served in the Republic of Korea Army since he graduated from the Korea Military Academy in 2004. After earning a Master's degree in the UAE, he completed a PKO mission in UNIFIL and a Defense Cooperation and Exchange mission in the UAE.

Part I

Introduction and Reflections on the Field

1 Introduction

Mission formations and a new agenda for the study of military units in action[1]

Eyal Ben-Ari, Uzi Ben-Shalom, Thomas Brond and Carmit Padan

Since the end of the Cold War, the world has seen the advent of what have variously been called the "New Wars" or "Hybrid Wars" (Hoffman, 2007; Kaldor, 2005; Munkler, 2005). These labels refer to conflicts combining differing forms of fighting (such as conventional, irregular or disruptive) in ways that blur their purportedly discrete nature. Thus, they may include a mixture of armed confrontations resembling conventional battles or skirmishes (pitting more or less regular forces against each other), anti-insurgency and policing campaigns, or efforts such as humanitarian assistance or reconstruction. Furthermore, they often encompass military and police units, civilian state organizations and multinational frameworks and non-state actors. For the armed forces, these kinds of conflicts imply that to older, conventional tasks have been added many new missions ranging between strictly military tasks, security-related assignments and civilian undertakings. What is striking about these missions, however, is that they are undertaken by specially created – and temporary – organizational formations, each of which is "tailored" to the missions at hand.

Yet, despite the advent of these new armed conflicts, much of the social scientific study of militaries-in-action seems to be "stuck" with tools developed in the heyday of conventional wars (dating back to the two World Wars and the Cold War). These social scientific tools focus on such classic issues as cohesion and leadership, communication and unit dynamics or discipline and motivation (Matthews, 2014; Schilling, 2019). Although all of these issues continue to be important, the problem is that almost all of these studies persist in focusing almost exclusively on organic units (up to and including brigades). In contrast, our volume suggests the utility of concepts related to mission formations – rather than "units" or "components" – to better capture the nature of the amalgamations and combinations that today's military missions involve. These new configurations link a variety of military units and a diverse set of governmental and non-governmental entities.

Yet these configurations – often temporary and constructed for specific missions – are analyzable as they are oriented towards goals, have internal structures and are marked by specific social and organizational characteristics. Our volume takes as its focus precisely these formations. Hence, it explores and develops new social scientific tools for the analysis and understanding of contemporary military action, that is, of armed forces *in theater*. Accordingly, it seeks to present and advance newer – or combinations of newer and older – analytical tools, methodologies and theories for the study of the military forces on deployment. For instance, whereas previous experiences of multi-nationality usually implied coordination at headquarters and the actual combat in uni-national formations, today even small units may find themselves interacting with other national units and civilian entities that may be characterized by different ethical codes, value systems and professional socialization.

To analyze these diverse mission formations – those relatively impermanent, modular frameworks constructed for specific missions – the contributions and the volume as a whole utilize innovative social scientific approaches developed over the past three or so decades. In other words, we show how the new wars provide empirical, methodological and theoretical opportunities for social scientists. In the rest of this introduction, we explain the empirical and analytical importance of mission formations, the older and newer social scientific tools that we draw upon and the main issues that are opened up for study and we introduce the chapters.

What are mission formations? Why study them?

By mission formations, we refer to combinations, fusions and blends of "tactical" units of the ground forces with a variety of specialized military forces and civilian entities in temporary, usually mission-specific amalgams within the context of contemporary conflicts. In formal military jargon, "tactical" refers to the level of war at which battles are planned and executed to accomplish objectives, and to forces organized to function in combat as self-contained entities. Because our aim is social scientific (and *not* doctrinal), however, we use this military term only as *shorthand* to underscore the kinds of structures and relations on which we focus – basically up to and including brigades. The sociological scientific study of these formations refers to the study of their composition, structures, interactions, processes and practices. We use sociology as an umbrella term referring to the disciplines – sociology, social psychology, anthropology, political science and organization science – that are relevant to the study of such mission formations.

The concept of mission formations – rather than "units", "elements" or "components" – is intended to capture the (ongoing) processual nature of the amalgamations, assemblages or combinations that military involvement in conflicts necessitates. These configurations are oriented towards goals and seek information about their environments, have internal (sometimes contradictory) structures and are marked by specific social and organizational characteristics and by different degrees of temporariness. In organizational parlance, they are action sets (Czarniewska, 2004, 2005). The concept of mission formations is used to underscore how militaries are becoming more akin to flexible organizations characterized by the relative loosening of internal and external boundaries.

Empirically, mission formations include relatively long-lasting forms such as peacekeeping frameworks (Autesserre, 2014), multinational expeditionary forces (van Fenema, 2009; Leonhard et al., 2008; Moelker et al., 2007; Shields, 2011;

Tomforde, 2009; Tresch, 2007; de Ward & Kramer, 2010) or frameworks created within the US Department of Homeland Defense (Fosher, 2008), temporary creations such as battle-groups (King, 2011), combined arms assault teams (Ben-Ari, 2015; King, 2010a), amalgamations of military units and police forces (Banks, 2016) or indeed much more ephemeral mergers as in an ad hoc combination of forces that join and rejoin an armed effort, structures composed for delivering food and medical aid or ad hoc joining of interpreters to organic forces (Hajjar, 2017; Van Dijk, Soeters & de Ridder, 2010). In addition, we would include organizational arrangements between armed forces and other state agencies, as well as non-governmental and international organizations (Osinga & Lindley-French, 2010, p. 24; Ben-Ari, 2018).

All these organizational forms contain fewer fixed structures and more temporary systems than "organic" units, and their constituent elements, people and technologies are assembled and disassembled according to the shifting needs of specific projects. Hence, mission formations include various forms of distributed teams or configurations of units from different arms and services or civilian agencies that are often put together in an ad hoc manner for specific tasks and missions, *including* violent encounters (Bollen & Soeters, 2010).

In focusing on mission formations, we go beyond an emphasis on multinational forms or jointness that still assumes relatively homogeneous, bounded ("textbook") units joined to other such units (Ruffa, 2018; Friesendorf, 2018). Rather, we emphasize the integration of diverse elements in a dynamic manner that may constantly change and that often include civilian components. To emphasize then, for our purposes, an infantry company calling in air support becomes, for the duration of the cooperation, a provisional combat formation. Similarly, an artillery officer and his signaller calling in direct fire have become a small team, coordinating a diversity of fire as the result of improved communication but for only a limited period of time (King, 2011, p. 255). More broadly, to provide another military example, we refer to such formations as the rapid reaction forces in contemporary Europe studied by Anthony King (2006, pp. 268–9) who notes that:

> They are no longer properly light infantry brigades, but have developed into hybrid, mobile brigades capable of maneuvering on a dispersed battlefield. At the same time, they are becoming joint organizations, with horizontal relations developing into supporting assets often from the air and maritime components. Finally, in light of the new operations, these brigades have recognized the centrality of new forms of intelligence and, indeed, information operations to the conduct of their missions.

In this view, in place of fronts, the "battlespace" consists of independent "lozenges" of discrete tactical activity: battle-groups – our mission formations – that are coordinated by a higher command may operate substantially independently of one another against threats which may come from any direction using distal but accurate fire power.

To be sure, such formations are not that new, but we suggest that, for analytical purposes, it may be fruitful to look at their dynamics rather than, as most sociology of military action has done, their constituent organic units. Theoretically, then, the crucial point of our formulation is that of a move from a sociology of combat units to a sociology of mission formations (in the dual sense of the word in English – as a noun and as a verb). If the platoon–company–battalion nexus was the focus of the older sociology (Moskos, 1984; Siebold, 2001), the new one focuses on combinations or assemblages of forces and entities for specific missions.

There are four reasons for focusing on this level of analysis: one historical and three analytical. First, from the perspective of the industrial democracies, the period after 9/11 has been one in which suddenly "everyone" was fighting or at least deployed to areas marked by armed conflict. For some militaries such as the American, British, French or Israeli armed forces, the conflicts were a continuation of previous armed engagements. But for many others, such as most of the European countries or Japan and South Korea, deploying troops to Iraq or Afghanistan (and other places) represented a first after the end of the Cold War. Thus, this period saw a significant enlargement of a "family" of fighting militaries participating in actual armed conflicts. But, because so many contemporary deployments are multinational and more often than not include civilian components, the armed forces have found themselves carrying out their assigned tasks in mission formations.

Second, analytically this is the level where the macro processes charted out by scholars – casualty aversion, marketization, technologization or juridification, to mention only a few – shape the use of organized state-sanctioned violence (the core of the military's expertise). Although the effects of these processes can be seen across the armed forces, in this volume we are interested in how they shape (and are shaped) at the level of militaries-in-use, in concrete operations. The idea is that, although there are continuities between the way the ground forces carry out their missions today and the way they have done so in the past, as new social environments have emerged, they have had a significant influence on the ways in which soldiers operate. Hence, if we want to chart how macro-sociological trends such as changed forms of legitimacy for using armed state violence, the social distribution of death or the impact of human rights on military action (Shaw, 2005; Levy, 2019) actually influence the level at which units perform, then this level is the most suitable. Similarly, the increasing concentration and internationalization of forces are long-term trends that need to be systematically analyzed at the level of mission formations (King, 2011).

Third, this level of analysis is characterized by forms of social organization and processes that are *analytically* distinct from higher levels. In other words, what goes on at this level is influenced by, but cannot be reduced to, external factors (Gazit & Ben-Ari, 2017). Macro approaches to the study of the armed forces do not suffice to analyze the kinds of interactions, emergent properties and logics of action that are found in mission formations because they tend to focus on the level of states and governing institutions, civil–military relations or the social origins of troops. In other words, whereas the macro social scientific approaches center on war or broad strategic issues, we are interested at the level of warfare or missions carried out at the "tactical" level.

Fourth, although the classic sociology of warfare has long provided tools for analyzing this level – in excellent studies of cohesion, leadership, communication or motivation (Moskos, 1971) – it is only relatively recently that the insights of new social scientific approaches to social life, to organizational dynamics and to violence have been applied to the military world. Hence, we suggest that using new social scientific tools and understandings developed over the past three or so decades may reinvigorate the study of combat, of combat units, of the new or renewed tasks they are assigned and especially of the new organizational forms that they operate within. Moreover, a focus on this level may prod us to develop new concepts for understanding military action and (no less important) direct us to think anew many of the classic questions of the sociology of combat.

The classic sociology of combat units

The key questions in the classic sociology of combat and combat units are centered on the contribution of different elements of social structure to the survivability and smooth operation of military units – basically, resilience and efficacy (Moskos, 1971, 1975; Schwartz & Marsh, 1999). This older sociology of tactical units developed during World War II and was honed during the subsequent decades through the Korean War, the Six-Day and Yom Kippur Wars that Israel waged and, to an extent, in Vietnam (Caforio, 2007; Ender, 2009; Janowitz, 1959, 1971; Kummel & Prufert, 2000; Moskos, 1971, 1975, 1988; Nadelson, 2005; Schwartz & Marsh, 1999; Siebold, 2007). The Cold War, pitting the forces of NATO against those of the Warsaw Pact, gave it further impetus. Hence, the questions that this sociology asked were what holds groups together within violent encounters? How do they go about achieving effective collective social action in the highly stressful circumstances of battle (King, 2014)?

Indeed, much of post-World War II sociology borrowed its governing theoretical armature from industry and emphasized the certainty created by control and the manipulation of variables to achieve predictability and surety (Ben-Ari et al., 2010). Along these lines, the model of conflict at the base of the classic sociology of combat units centered on a conventional interstate conflict between regulars, fought in accordance with codified laws of war. The guiding assumptions in much of its imagery entailed clearly defined opponents, plain lines of territorial domination, quantifiable progression in war, and unambiguous links between military goals and the means to achieve them (Ben-Ari et al., 2010). Moreover, these images were accompanied by strong suppositions about what military structures "look" and "act" like, that is, wars are fought by "textbook units" that are based on hierarchy, ordered decision-making, internal consistency between organizational levels, and the availability of resources. This is a vision of combat units as squads, platoons, companies or battalions (and so on) – as found in popular imagery, scholarly portrayals and professional depictions. Violence was organized bureaucratically through units and subunits down to individuals who operated according to rules, regulations and set maneuvers within highly disciplined groups. Ideally, the actual violent act – shooting, hitting, killing, wounding – was to be directed hierarchically and according to specific orders based on clear objectives.

The reasoning behind the study of textbook units involved the combat environment of risk and violence that necessitated a special kind of socialization of soldiers and the establishment of strong arrangements for social support. The overall ideas were that of a machine that withstands the friction of battle (or disintegrates given enough pressure) and that of units as organically based on strong ties between individuals (Ben-Ari, 1998). In other words, interpersonal ties were to be cultivated to created resilience (group and individual) in the face of violence (Bartone and Hystad, 2010). Unsurprisingly, the epitome of this point is expressed in the literature on both cohesion and combat effectiveness (Sarkesian, 1980). The ensuing research thus included studies of socialization, discipline, motivation, cohesion, leadership and deviance, and within this sociology individuals were to be motivated, socialized, strengthened, disciplined, trained and led. This classic sociology of combat units was often used for prescriptive purposes as it basically offered recipes for creating individuals and units that could perform military tasks within the stressful conditions of conventional combat. These formulas included prescriptions for "creating cohesion", "putting in place leadership" or "encouraging confidence in one's weapons". Theoretically, there were strong functionalist assumptions (common to more general strands in the psychology and sociology of the time) at the base of analyses centered on the idea that different roles and attributes all contributed to (or deviated from) proper functioning.

Put simply, but not simplistically, this kind of sociology seems ill-suited for tackling the emergent, boundary-spanning, simultaneous nature of the military amalgams participating in many contemporary conflicts. Indeed, it appears that this classic sociology seems to have come to a standstill and has not offered significant new analytical concepts or theoretical formulations beyond those developed in the period between World War II and the end of the Cold War. Yet a careful look at current scholarship placed at the mezzo and micro sociological levels reveals a host of suggestions and intimations regarding what we see as an emerging agenda for the sociology of mission formations (Ben-Shalom et al., 2019; King, 2006, 2010b; Ouellet, 2005; Soeters, van Fenema & Beeres, 2010b; de Ward & Soeters, 2007). It is precisely this agenda – not always coherent or explicit, nor fully based on empirical research – that this introduction and this volume seek to chart (also Ben-Ari, 2016). At the risk of immodesty, we aim to explicate the theories and examples that have been mostly hidden, neglected or minimized because of the limiting analytical lenses of previous theories.

A new formulation

It is thus to the sociological and organizational – *not* technological or doctrinal – significance of the social order of mission formations that our agenda, and this book, is dedicated. To be crystal clear, we do not argue for abandoning established traditions of research, but rather for opening and adding to them new perspectives and questions. For example, cohesion, leadership and small group dynamics continue to be important in the new wars and in the emerging organizational forms through which they are pursued, but they are socially organized and organizationally significant in new ways. Or, take the most basic unit of the classic sociology of the armed forces: the squad or platoon (and sometimes, the company). In conventional circumstances, such units were usually trained to fight as cohesive forces – with clear boundaries and internal coherence – until their "staying-power" weakened or disintegrated. Today, however, a company or a battalion may sometimes be likened to a sort of holding company whose constituent units can be divided and joined to other pieces in new, often improvised, entities.

At the most basic, and perhaps trivial, level, a contemporary sociology of mission formations needs to be directly linked to the daily doings – the interactions, practices and emergent structures – of units and frameworks, to their actual operational dynamics (Soeters, 2018). As stated earlier, when forces from the industrial democracies came under the direct observation or proximal observation of social scientists, then it began to be clear that the classic issues of hierarchy, cohesion, leadership, motivation or discipline did not suffice to explain their actions and missions.

What, then, should a contemporary sociology of mission formations look like?

Broadly put, the move is from the older sociology of combat units developed for state-on-state wars to an analysis of collective action in mission formations within long-term, complex conflicts and within changed social expectations about how warfare is to be waged. The actions of these formations involve missions centered on organized military violence and an increasing multiplicity of other tasks. Finally, individuals are moved by older and newer bases of motivation, develop new capacities in addition to older ones, belong to groups that are intermittently created and recreated and serve under tactical leaders who must manage local-level complexity, translate strategic aims to doable actions and engender followership.

As the chapters in this volume deal with many of the issues set out in this formulation, in the rest of the introduction we chart a number of key themes that are interwoven throughout the book.

Mission formations and environments

Environments shape the forms and practices through which formations' constituent components achieve concerted, coordinated collective action. A variety of social, political and economic transformations have come to shape the wielding of state-mandated armed force. The armed forces, from their point of view, must assure continued legitimacy, that is, meet external expectations, convince outside publics, create favorable images of themselves and actively construct their necessity to continue receiving support and resources. Thus, militaries must constantly and actively manage their relations with their social, economic and political environments.

In this respect, perhaps the most important shapers of tactical-level operations are changed expectations related to demands for restrained and judicious use of military force. The major expression of this emphasis is, of course, the minimization of "unnecessary casualties", both on our side and among bystanders (Ben-Ari, 2005; Levy, 2019). Derived from a combination of ideas about human rights and the greater precision of contemporary weapons, militaries are expected to reduce as much as possible what in military jargon is called "collateral damage". What seems to have happened is the advent of what Shaw (2005) calls "risk-transfer war", centering on minimizing life risks to the military and thus the political risks to their civilian leaders. This point has meant that an assessment of risk of casualties has become a normal part of considerations at the level of tactical mission formations, and that military lawyers have been increasingly integrated into tactical decision-making (Cohen & Ben-Ari, 2014; Dickinson, 2010; Forster, 2012). From our perspective, this means that small units of legal experts have now been integrated into the combat formations used to deploy military force. In a closely related manner, legal expectations have been accompanied by changed ideas about whether and how military forces provide various kinds of aid and humanitarian assistance. As a consequence, members of the military have found themselves liaising or working directly with NGOs, local government or tribal structures (Byman, 2001; Ben-Ari, 2018; Winslow, 2002) in temporary, usually ad hoc, formations.

These developments, in turn, are related to what Shaw (2005) calls the appearance of "global surveillance": the growing transparency of contemporary armed forces to external agents such as political leaders, the media (local and global), the judiciary, pressure groups or international non-state institutions. This trend implies that troops are constantly open to external monitoring. In fact, as technological developments have made any person with a camera, computer or internet link a potential reporter, and troops have ready access to a plethora of news outlets, media reports have become very important for soldiers on the ground. Accordingly, the armed forces have developed various organizational appendages at the local level – such as media liaisons or press officers – to control information and present the more positive aspects of military actions (Shavit, 2017). This situation has meant the further addition of military media experts into operational formations.

Among the changed expectations about the activities of the military as an organization is privatization centered on demands by civilian leaders for leaner, less expensive forces based on privatized logistics and commercial companies supplanting state forces (Bury, 2019). As Berndtsson (2009) and Kinsey (2014) note, the use of private security forces has expanded since the end of the Cold War to spread beyond administrative and logistical roles to the battlefield and to related roles such as defensive guarding, security sector reform and disarmament. This situation implies, again, that military units sometimes find themselves within organizational formations that include civilian contractors that provide not only logistics, but at times also security.

Generally speaking, then, these examples attest to how environments are shaping the very content and dynamics of today's mission formations. And this situation in turn has not only created greater potential for accomplishing missions, but also increasing complexity and the need for more and more mechanisms to bridge between the different cultures, practices and expectations of diverse groups (Holmes-Eber, 2014; Rubinstein, 2014).

Individuals, motivations and capacities

In general, troops are involved in many of the same issues and face many similar challenges that the older sociology of combat units investigated such as resilience, gaining confidence in their weapons, finding their place in cohesive groups or developing leadership skills. But they increasingly find themselves needing new skills and looking for different incentives from military service. Take motivations. Battistelli and his colleagues (1997), examining Italian peacekeepers, found that in addition to older bases for motivation (commitment to the country) many soldiers evinced what they call post-modern motivations (looking to "do good"). The presence of multiple motivations and soldierly skills can be problematic, however, as in the processes of "training down" from combat to other missions, as the particular control and discipline required of soldiers and selected in them can lead to difficulties in using them in policing missions or in cooperating with civilian entities. In addition, in the more combat-oriented soldierly professions, prestige continues to be based on the distance from or nearness to struggles resembling conventional war. Hence, working with NGOs to distribute food is considered to be less interesting and of a lower standing than going out on armed patrols.

What is clear, however, is that, to operate in the new mission formations, soldiers are developing new capacities such as abilities to seek information and resources from civilian entities or behaving not only according to purely military norms and rules, but according to expectations that they be aware of the wider implications of their actions, as in the strategic corporal or strategic private (Schmidtschen in Shields, 2011, p. 21). Moreover, within these formations, individuals continue to belong to small groups but are intermittently joined to other forms that are created and recreated. Thus, it seems that the ability to join and disengage with other units and entities has become one of the capacities possessed by "good" soldiers and officers.

Leadership

Along the lines of our general argument, leadership in contemporary missions necessitates a model that is additive rather than linearly transformative: in other words, this model underscores the co-existence of older, more conventional values and orientations (Ben-Ari, 2011; Holenweger, Jager & Kernic, 2017) alongside (and not necessarily being replaced with) newer emphases in leadership (King, 2019). Consequently, tactical military leaders continue to need at least the psychological and physical levels of their subordinates, along with initiative and courage. But such successful leaders are also marked by new capacities such as the ability to be in charge of organic structures *and* loosely coupled systems; to use changed bases of power, building consensus *in addition to* formal authority; to lead under constantly shifting contingencies and spanning across national and organizational boundaries; and to provide guidance in ideologically and politically fraught circumstances (Shamir & Ben-Ari, 2000, 2009; Ben-Ari, 2011).

More broadly, within the evolving projects that mission formations engage in, leaders are parts of "open systems". Indeed, because present-day militaries have developed new forms of cooperation with components drawn from a variety of forces or external organizational entities, each with its own doctrines, traditions and modes of operation, many military leaders are mediators between inside and outside their own units, connectors between the diverse entities of the formations and guides in regard to the emergent ventures. Moreover, given the political complexity and at times volatility of some conflicts, leaders often find themselves also translating strategic aims into tactical prescriptions (Ben-Ari, 2018) or constantly negotiating with partners having different agendas and interests (Caforio, 2009). The centrifugal pulls of mission formations (often loosely coupled, marked by fragmented cultures and possessing new technologies) mean, moreover, that commanders have become "centers of gravity" that use a variety of bases for their authority, including relationship building, persuading, framing reality or consensus building, in addition to the more conventional ones of hierarchical authority and rank. More widely, then, in such circumstances a primary task of leaders is to provide the mental models and frames to coordinate the behavior of organizational members and to provide purpose and meaning (Shamir & Ben-Ari, 2000, 2009).

Introducing the chapters

The organization of the volume's chapters involves a move from the broad theoretical, analytical and methodological issues involved in the study of mission formations to the organizational issues and challenges posed by their creation. The idea is to provide a basis for the study of such configurations and, thus, an analytical background to the specific studies.

The chapter by Eric Ouellet, "New directions in military sociology: reflections on a book project 15 years later", complements this introduction. In this contribution, Ouellet argues that military sociology has been very much focused on military personnel, with an emphasis on tactical issues such as cohesion, morale and small unit leadership or larger personnel policies such as recruitment, retention and training. These studies tend to be very much functionalist in nature and, thus, take for granted many institutional assumptions that in the 21st century are highly questionable: for instance, about the meaning of military and political objectives. As a consequence, Ouellet explains that he decided 20 years ago to edit an anthology to show how non-functionalist perspectives could enrich the study of military affairs; it was eventually published in 2005 and is entitled *New Directions in Military Sociology*.

Developing his thinking and the themes raised in this volume, he takes up a non-functionalist approach with an emphasis on the much-neglected sociological study of the operational level of war where military planning and decision-making take place, and where the fate of mission formations in theater is set in motion. He shows that the implicit shared assumptions of military planners, be they cognitive or normative, can have a substantive impact on how combat formations are used, for better or for worse. Hence, he suggests that a fruitful way to further research is not to discover behavioral patterns in planning to maximize military efforts, but to develop methodologies to uncover deep patterns in shared military thinking, to identify poorly thought through shared assumptions. His chapter provides a review and assessment of ongoing work on institutional analysis, campaign design and three NATO projects where he was involved. In terms of our volume, this chapter places mission formations squarely within their social and political environments.

The chapters that follow are arranged around three clusters of issues: the organizational forms characterizing mission formations, methodological challenges in studying such forms, and the particular dynamics of allied amalgams that are both global and local, followed by an epilogue. The part on the organization of mission formations begins with Wilbur Scott's "Organizational adaptations in the hunt for Abu Musab al-Zarqawi: a review of concepts for analyzing their usefulness", which takes up another type of formation. In 2004, General (retd) Stanley McChrystal, commander of Joint Special Operations Task Force (SOTF), Iraq, evaluated his group's readiness to take on a violent, upstart terrorist organization, al-Qaeda in Iraq (AQI). At the time, the SOTF was a textbook model of efficiency. However, though trained and equipped at inferior levels, AQI's tactical agility in carrying out terrorist and disruptive acts exceeded SOTF's ability to stop or even interrupt them. AQI's modus operandi, McChrystal concluded, was better suited to the contingencies of a 21st-century complex battlefield in which flexibility and adaptability are more important than efficiency.

McChrystal's solution was to reorganize SOTF as a team of teams: small special operations teams, given the discretion to carry out the commander's intent, were networked with teams in other sectors of the organization also authorized to carry out their own specialized pieces of action. This enabled cross-silo communication and information flows, enhanced common purpose and shared awareness, and empowered time-sensitive action (Goldenberg & Dean, 2017). These days, McChrystal-esque tactical operations centers (TOCs) interconnected with joint, task-specific teams are a prime mission formation in the US military's down-range organizational arsenal. However, from a theoretical point of view, explaining why such configurations work or not requires different concepts than those usually employed in traditional sociological analyses of combat units. This chapter documents McChrystal's adaptations and reviews concepts for assessing their usefulness for military action in complex situations. Scott's contribution underlines the processual nature, analytically self-organizing or better still self-forming, of mission formations.

Jessica Glicken Turnley's chapter, "Bureaucracies, networks and warfare in a fluid operating environment", explains that in recent decades discussions about formal organizational structures such as bureaucracies have become increasingly outpaced by debates about network forms. Social networks' quick-response capabilities make them well suited for a rapidly changing world. Accordingly, Turnley asks: to what extent has – or perhaps, should – the military parallel this shift from bureaucracies to networks, using relationships to draw appropriate resources into a crisis situation?

Joint task forces (JTFs) – a distinct form of mission formation – are one of the US military's primary vehicles for operational, mission-focused engagement across military services. This chapter argues that hierarchical (bureaucratic) structures within JTFs continue to provide value, although there may be a space for mission-driven networks in their operation. An American civilian management and coordination structure called the Incident Command System (ICS) is used to illustrate how both networks and hierarchic structures are always at play together in rapid response situations. Using examples from the ICS and historic special operations forces incidents, it demonstrates that the development of strong interpersonal relationships during non-crisis periods creates networks and shared representations that can be leveraged by formal structures during crisis situations. The wider point her chapter makes is that the challenge lies in the ability of relevant organizations to support and promote cross-organizational personal engagement without turning it into a formal bureaucratic requirement.

Eyal Ben-Ari's contribution, "From leading combat units to leading combat formations: modularity, loose systems, and temporariness", rounds off this section. He takes up the question of "What does command look like in mission formations?" The question goes beyond accepted questions about command of organic units to ask about the special character of leading temporary, often ad hoc, military configurations established for specific missions. Following King's (2019) suggestions about command in the 21st century, he uses the two volumes published by the Swedish Centre for Studies of Armed Forces and Society (Tillberg & Tillberg, 2013; Tillberg et al., 2017) that document dialogues with military personnel who had been deployed outside the country since the end of the Cold War.

Ben-Ari contends that, although the older armature of military sociology continues to be useful in the analysis of such formations, there is a need to supplement it with concepts derived from contemporary organizational science that takes into account the new kinds of organizational forms that are used in a variety of fields. Thus, although classic issues such as motivation, leadership, cohesion or bureaucratic procedures continue to be important, to understand mission formations it is especially important to analyze such things as the creation of a sense of common goal or fate among diverse actors, how coordinated collective action is achieved using not only military authority, but also persuasion and negotiation, and how trust is engendered in temporary configurations. Central to these efforts is the emergence of informal contracts or understandings between the components of the mission formations.

The next part of the volume – Methodologies for the Study of Military Formations – comprises one contribution by Uzi Ben-Shalom, which is entitled "Research approaches to the study of combat formations: a personal note". Ben-Shalom contends that the study of mission formations involves all of the "usual" difficulties social scientists encounter in studying the armed forces, such as problems of access to units and operations, work with internal military researchers or achieving critical distance from informants (Ben-Ari & Levy, 2014; Carreiras & Castro, 2012). Particular to mission formations, however, are the difficulties of designing a set of methods and practices that allow scholars to study the diversity of the frameworks, the processes by which they are created (and disbanded) and their internal dynamics. Ben-Shalom exemplifies these considerations through four empirical studies that he carried out with colleagues about the IDF.

His conclusion is that the best research design includes researchers possessing both "native knowledge" (that is, understanding of professional military considerations) and a diversity of academic expertise. Moreover, conceding that access to operations while they are ongoing may be very difficult (if not impossible), Ben-Shalom recommends thorough post-mission studies. This kind of design is also important as the very character of many mission formations as emergent projects implies that it is only post-factum that their aims, successes (and failures) and internal processes and interactions are clarified. Thus, flexibility in design, as well as thoroughly sourcing data from a large variety of sources, appears to be the best kind of solution to the problems of studying mission formations.

The third cluster of articles – Glocalized Mission Formations – centers on the kind of relations that emerge within mission formations that cross national lines, a type of configuration that has almost become the default model for the armed forces of the democracies. Insoo Kim and Young-Il Choi's chapter, "Institutional isomorphic change in South Korea's UNPKO mission formation", echoes many of the issues raised earlier by Ouellet. As they describe it, the end of the Cold War in the late 1980s has brought about a major change in the missions assigned to the South Korean armed forces. As the ideological East–West conflict broke up, the two Koreas could join the United Nations at the same time. Soon after South Korea joined the United Nations in 1991, the South Korean government was asked to send South Korean troops to carry out a UNPKO mission. Since then, at the request of the UN, South Korea has participated in seven UNPKO activities since 1993.

However, because the South Korean military was not prepared to conduct peacekeeping operations within the context of multinational mission formations in the 1990s, it had to develop a way of carrying out sets of relevant practices such as providing medical care or engineering expertise. What emerged are interesting features in the PKO mission formations. First, PKO units did not have a fixed organizational form, but were temporarily organized to carry out missions assigned under a UN mandate. Second, the PKO units consisted of action sets in which infantry, engineering and medical officers worked in conjunction with one another with some degree of regularity. Third, the PKO units reflected not a rational choice for greater efficiency, but rather norms and values that justified their activities. Kim and Choi's study argues that the consistency of PKO mission formation is the product of an institutional environment that refers to a socially constructed normative world. In other words, they show how the structure of specific mission formations is shaped not (just) by considerations of effectiveness, but by political and organizational expectations in their environments.

Thomas Randrup Pedersen's contribution is entitled "'Democracy … 120 mm at a time': mission formations and operational entrapments in post-9/11 Afghanistan". Pedersen explores mission formations within the context of the NATO-led International Security Assistance Force (ISAF). More specifically, his chapter examines Danish ISAF troops as part of two multinational mission formations: the UK-led Task Force Helmand and the US-led Task Force Belleau Wood. Based on ethnographic fieldwork in theater, his chapter zooms in on some of the very last Danish combat units deployed to Helmand. Striving for a novel insight into contemporary military action through a "view from within", it draws upon the anthropology of morality/ethics to address the classic issue of soldiering and motivation.

What Pedersen shows is that, on the basis of the Danish troops' dual role as both "soldier-diplomats" and "soldier-warriors", soldiering can be conceptualized as a project of self-becoming along two lines: a "force for good" and a "force of death and destruction". He probes into ways in which such potentially conflicting projects of soldierly selves make themselves manifest in military action. Further, he argues that the Danish rank and file soldiers, with whom he was embedded, are concerned not so much with making the world into "a better place" as with making themselves into "greater warriors". He further argues that soldierly becomings embody a desire for proving one's worth to oneself and to others, *including* one's allies. He shows that this sense of worth is situated within a field of tension between notions of potency and professionalism. What this chapter underscores are the dynamics of constant comparison with warriors of allied forces (in this case the US and British troops) in constructing their soldierly selves. These dynamics, often implicit, govern many of the informal relations that emerge in mission formations.

The chapter by Joseph Soeters, Gerold de Gooijer, Paul C. van Fenema and Nuno Oliveira, "Logics battlefield: IT contracting and military reserves in the Dutch Army", focuses on cooperation between the armed forces on the one hand and both reservists and civilian contractors on the other in the realm of information technology. These two sorts of actors are often seen by military commanders as a sort of flexible layer of "employees" or "laborers" around the (combat) core. Such a layer is necessitated given the demands on military formations to be constantly adaptive to changing circumstances. As they explain, today militaries include not only multiple organizational identities (e.g., combat soldiers and peacekeepers) and multiple organizational forms (e.g., strict hierarchies and ad hoc structures), but also multiple logics of action: the assumptions, interpretive frameworks, prescriptions and practices held together by an internal coherence (a "fit") that are seen as reasonable and doable by actors in certain, analytically limited spheres of action (Jackall, 1988, p. 112; Thornton & Ocasio, 2008: 10; Eran-Jona & Ben-Ari, 2014).

Their contribution centers on the Netherlands' military, an organization consciously endeavoring to become more adaptive through interorganizational collaboration with other organizations. Specifically, it analyzes the collaboration of the national military with civilian hospitals, ICT companies and reservists as a special component of the workforce not belonging to the core of the organization itself. Their case study thus focuses on managing organizational hybrids and the encounter between a military logic and external logics of both for-profit businesses and reservists who are both within and outside the military. The adaptive value of such linkages involves the use, by the armed forces, of the specific skills, knowledge, creativity and innovation power brought in by outsiders. The authors' theoretical proposals, based on notions of hybrid organizing, center on the meeting of different logics of action. They suggest that there are three problems involved in such organizing: integration, differentiation and combination.

The epilogue, by Thomas Crosbie, both completes the examination of the macro level and provides another integrative voice to the volume as a whole. It seeks to link the sociological and organizational analyses contained in the volume with military thinking about today's missions and the forms through which they are pursued.

Conclusion: towards a new sociology of mission formations

The various chapters and the volume as a whole underscore the need for a much more dynamic social and organizational view of militaries-in-use than that provided by classic formulations in the sociology of combat units. The move is from seeing the military as an entity that has to withstand the pressures of missions (ultimately battle) to an appreciation of the contingent, often contradictory, organizing processes that constantly take place within mission formations. In other words, the shift we suggest is from a focus on the resilience of organic units to the questions about how collective action is achieved under conditions of impermanent multi-actor formations.

To reiterate, however, we do not argue for abandoning established traditions of research, but rather for opening them up to new ways of looking at things and introducing new perspectives. In this manner, and in contrast to some recent tomes often (over)emphasizing revolutions or transformations in the way war is waged, conventional warfare continues to be germane to contemporary scholarship in ways that open up rather than limit our understandings.

Thus, for instance, the innovative approaches suggested in this volume take into account how classic military socialization continues to color peacekeeping and policing, and how impelling soldiers to action involves both older and newer bases of motivation. To repeat, many of the forms and processes that we have dealt with have historical antecedents, but the combination of many of the attributes of today's conflicts, as well as developments within the social sciences themselves, have brought us to a point where scholars may be ready for new concepts and questions about militaries-in-use. Thus, we argue for a model that is cumulative rather than linear, in the sense that new military capacities, conditions and organizational configurations have been added to conventional ones. Indeed, organic units continue to be important as they are the frames around which many mission formations are constructed.

Finally, the study of mission formations must always take into account the actual or potential use of armed violence. The social organization of such violence may contain, restrain, direct, control, buffer, utilize, coordinate, distribute, allocate and orchestrate it, but we underscore the need for a multiplicity of verbs (a repertoire of verbs) to understand how missions are carried out at the tactical level. If in the (still relevant) older sociology the primary emphasis was on the effectiveness of units and the resilience of individuals, small groups and units in the face of violence, today the idea is to look at a wider selection of issues for analysis.

Note

1 We would like to thank participants in a previous workshop (funded by the Dan Shomron Kinneret Center for Society, Security and Peace and Kinneret Academic College) held in the Jerusalem hills in September 2017, and especially Eric Ouellet and Kobi Michael for their comments on this agenda, and participants at the famous Gniben Seminar held on a Danish peninsula in June 2018. We would like to thank the Royal Danish Defence College for funding and hosting this seminar.

References

Autesserre, S. (2014). *Peaceland: Conflict resolution and the everyday politics of international intervention*. Cambridge: Cambridge University Press.

Banks, Wa.C. (2016). *Soldiers on the home front: The domestic role of the American military*. Cambridge, MA: Harvard University Press.

Bartone, P., & Hystad, S.W. (2010). Increasing mental hardiness for stress resilience in operational settings. In P. Bartone, B.H. Johnsen, J. Eid, J.M. Violanti & J.C. Laberg (Eds.), *Enhancing human performance in security operations* (pp. 257–272). Springfield, IL: Charles C. Thomas.

Battistelli, F. (1997). Peacekeeping and the postmodern soldier. *Armed Forces and Society, 25*(3), 467–483.

Ben-Ari, E. (1998). *Mastering soldiers*. New York: Berghahn Books.

Ben-Ari, E. (2005). A "good" military death: Cultural scripts, organizational experts and contemporary armed forces. *Armed Forces and Society, 31*(4), 651–664.

Ben-Ari, E. (2011). Challenges of leadership in contemporary expeditionary operations. In H. Furst & G. Kummel (Eds.), *Core values and the expeditionary mindset: Armed forces in metamorphosis* (pp. 35–50) Berlin: Nomos.

Ben-Ari, E. (2015). From a sociology of units to a sociology of combat formations: Militaries and cohesion in urban combat. In A. King (Ed.), *Frontline: Combat and cohesion in the twenty-first century* (pp. 73–92). Oxford: Oxford University Press.

Ben-Ari, E. (2016). What is worthy of study about the military? The sociology of militaries-in-use in current-day conflicts. In H. Carreiras, C. Castro & S. Frederic (Eds.), *Researching the military*. Routledge.

Ben-Ari, E. (2018). From leading combat units to leading combat formations – Modularity, loose-systems and temporariness. In L.V. Tillberg (Ed.), *Uppdrag militar* (pp. 53–70). Stockholm: Center for Society and Military Studies.

Ben-Ari, E., Lehrer, Z., Ben-Shalom, U., & Vainer, A. (2010). *Rethinking the sociology of combat: Israel's combat units in the Al-Aqsa intifada*. Albany, NY: State University of New York Press.

Ben-Ari, E., & Levy, Y. (2014). Insider/outsider perspectives. In J. Soeters & P.M. Shields (Eds.), *Routledge handbook of research methods in military studies* (pp. 9–18). London: Routledge.

Ben-Shalom, U., Lewin, E., & Engel, S. (2019). Organizational processes and gender integration in operational military units – An IDF case study. *Gender Work and Organization*, *26*(9), 1289–1303.

Berndtsson, J. (2009). *The privatization of security and state control of force: Changes, challenges and the case of Iraq* (PhD dissertation) School of Global Studies, University of Gothenburg.

Boene, B. (1990). How unique should the military be? A review of representative literature and outline of synthetic formulation. *Archives Européennes de Sociologie*, *31*(1), 3–59.

Bollen, M., & Soeters, J. (2010). Partnering with "strangers". In J. Soeters, P.C. van Fenema & R. Beers (Eds.), *Managing military organizations: Theory and practice*. London: Routledge.

Bury, P. (2019). *Mission improbable: The transformation of the British Army Reserves*. Hovant, Hampshire: Howgate.

Byman, D.L. (2001). Uncertain partners: NGOs and the military. *Survival*, *43*(2), 97–114.

Caforio, G. (2007). Introduction: The interdisciplinary and cross-national character of social studies on the military – The need for such an approach. In G. Caforio (Ed.), *Social sciences and the military: An interdisciplinary overview* (pp. 1–20). London: Routledge.

Caforio, G. (2009). Rhetorical persuasion and storytelling in the military. In G. Kummel, G. Caforio & C. Dandeker, *Armed forces, soldiers and civil–military relations*. VS Verlag fur Sozialwissenschaften.

Carreiras, H., & Castro, C. (Eds.). (2012). *Qualitative methods in military studies: Research experiences and challenges*. Routledge.

Cohen, A., & Ben-Ari, E. (2014). Legal-advisors in the armed forces: Military lawyers in the Israeli Defence Forces as mediators, interpreters and arbitrators of meaning during operations. *Journal of Political and Military Sociology*, *42*, 125–148.

Czarniawska, B. (2004). *On time, space and action nets* (Gothenburg Research Institute Report 5) Gothenburg University.

Czarniewska, B. (2005). Networks, networking and nets: NCW from an organization theory perspective. In K. Yden (Ed.), *Directions in military organizing*. Stockholm: Forvarshogskolan.

Dickinson, L.A. (2010). Military lawyers on the battlefield: An empirical account of international law compliance. *American Journal of International Law*, *104*(1), 1–28.

Ender, M.R. (2009). *American soldiers in Iraq: McSoldiers or innovative professionals*. Routledge.

Eran-Jona, Meytal, & Ben-Ari, Eyal (2014). Hybrid conflicts, multiple logics and organizational transitions: Military relations with local civilians. *Res Militaris*, *4*(2), 1–14.

Forster, A. (2012). The military covenant and British civil–military relations: Letting the genie out of the bottle. *Armed Forces and Society*, *38*(2), 273–290.

Fosher, K. (2008). *Under construction: Making homeland security at the local level*. Chicago, IL: University of Chicago Press.

Friesendorf, C. (2018). *How western soldiers fight: Organizational routines in multinational missions*. Cambridge: Cambridge University Press.

Gazit, N., & Ben-Ari, E. (2017). Military violence in its own right: The micro-social foundations of physical military violence in non-combat situations. *Conflict and Society: Advances in Research, 3*, 189–207.

Goldenberg, I., & Dean, W.H. (2017). Enablers and barriers to information sharing in military and security operations: Lessons learned. In J. Soeters, I. Goldenberg & W.H. Dean (Eds.), *Information sharing in military operations* (pp. 251–268). Berlin: Springer.

Hajjar, R. (2017). Effectively working with military linguists: Vital cultural intermediaries. *Armed Forces and Society, 43*(1), 92–113.

Hoffman, F.G. (2007). *Conflict in the 21st century: The rise of hybrid wars.* Washington, DC: Potomac Institute for Policy Studies.

Holenweger, M., Jager, M.K., & Kernic, F. (Eds.). (2017). *Leadership in extreme situations.* Springer.

Holmes-Eber, P. (2014). *Culture in conflict: Irregular warfare, culture policy and the Marine Corps.* Stanford, CA: Stanford University Press.

Jackall, R. (1988). *Moral mazes. The world of corporate managers.* Oxford: Oxford University Press.

Janowitz, M. (1959). *Sociology and the military establishment.* New York: Russel Sage Foundation.

Janowitz, M. (1971). *The professional soldier: A social and political portrait.* New York: Free Press.

Kaldor, M. (2005). *New and old wars: Organized violence in a global era.* London:Polity.

King, A. (2006). The post-Fordist military. *Journal of Political and Military Sociology, 34*(2), 359–374.

King, A. (2010a). The internationalization of the armed forces. In J. Soeters, P.C. van Fenema & R. Beeres (Eds.), *Managing military organizations* (pp. 42–54). London: Routledge.

King, A. (2010b). Understanding the Helmand campaign: British military operations in Afghanistan. *International Affairs, 86*(2), 311–332.

King, A. (2011). *The transformation of Europe's armed forces: From the Rhine to Afghanistan.* Cambridge: Cambridge University Press.

King, A. (2014). *The combat soldier.* Oxford: Oxford University Press.

King, A. (2019). *The twenty-first century general.* Cambridge: Cambridge University Press.

Kinsey, C. (2014). Transforming war supply: Considerations and rationales behind contractor support to UK overseas military operations in the twenty-first century. *International Journal, 69*(4), 494–509.

Kummel, G., & Prufert, A.D. (Eds.). (2000). *Military sociology: The richness of a discipline* (pp. 379–406). Frankfurt: Nomos.

Leonhard, N.G., Aubry, M., Casas, S., & Janowoski, B. (2008). *Military cooperation in multinational missions: The case of EUFOR in Bosnia and Herzegovina.* Socialawissenshcaftliches Institut fur der Bundeswehr.

Levy, Y. (2019). *Whose life is worth more? Hierarchies of risk and death in contemporary wars.* Stanford: Stanford University Press.

Matthews, M. (2014). *Head strong: How psychology is revolutionizing war.* Oxford: Oxford University Press.

Moelker, R., Soeters, J.M.M.L., & vom Hagen, U. (2007). Sympathy, the cement of interoperability. *Armed Forces and Society, 33*(4), 496–517.

Moskos, C. (1971). *The American enlisted man: The rank and file in today's military*. New York: Russell Sage Foundation.

Moskos, C. (1975). The American combat soldier in Vietnam. *Journal of Social Issues, 31*(4), 25–37.

Moskos, C. (1988). *Soldiers and sociology*. Washington, DC: United States Army Research Institute for the Behavioral and Social Sciences.

Moskos, C. (1984). Review: The sociology of combat. *Contemporary Sociology, 13*(4), 420–422.

Munkler, H. (2005). *New wars*. Polity.

Nadelson, T. (2005). *Trained to kill: Soldiers at war*. Baltimore, MD: Johns Hopkins University Press.

Osinga, F., & Lindley-French, J. (2010). Leading military organizations in the risk society: Mapping the new strategic complexity. In J. Soeters, P.C. van Fenema & R. Beeres (Eds.), *Managing military organizations, Theory and practice* (pp. 17–28). London: Routledge.

Ouellet, E. (Ed.). (2005). *New directions in military sociology*. North York, ON: De Sitter.

Rubinstein, R.A. (2014). Humanitarian–military collaboration: Social and cultural aspects of interoperability. In R. Rubinstein (Ed.), *Cultural awareness in the military: Developments and implications for future humanitarian cooperation* (pp. 57–72). New York: Palgrave Macmillan.

Rubinstein, R.A., & Zoli, C. (2019). Civil–military relations from international conflict zones to the United States: Notes on mutual discontents and disruptive logics. In B.R. Sørensen & E. Ben-Ari (Eds.), *Civil–military entanglements*. Oxford: Berghahn.

Ruffa, C. (2018). *Military cultures in peace and stability operations: Afghanistan and Lebanon*. Philadelphia, PA: University of Pennsylvania Press.

Sarkesian, S. (Ed.). (1980). *Combat effectiveness: Cohesion, stress and the volunteer army*. New York: Sage.

Schilling, S. (2019). *Cohesion in modern military formations: A qualitative analysis of group formation in junior specialized and ad-hoc teams in the Royal Marines* (Doctoral dissertation). School of Security Studies, King's College London.

Schwartz, T.P., & Marsh, R.M. (1999). The American soldier studies of WWII: A 50th anniversary commemorative. *Journal of Political and Military Sociology, 27*(1), 21–37.

Shamir, B., & Ben-Ari, E. (2000). Challenges of military leadership in changing armies. *Journal of Political and Military Sociology, 28*(1), 43–59.

Shamir, B., & Ben-Ari, E. (2009). Hybrid wars, complex environments and transformed forces: Leadership in contemporary armed forces. In A.J. Gilie van Dyk (Ed.), *Strategic challenges for the African armed forces for the next decade* (pp. 1–16). Pretoria: Sun Press.

Shavit, M. (2017). *Media strategy and military operations in the 21st century*. Routledge.

Shaw, M. (2005). *The new western way of war*. London: Polity Press.

Shields, P. (2011). An American perspective on 21st century expeditionary mindset and core values: A review of the literature. In H. Furst & G. Kummel (Eds.), *Core values and the expeditionary mindset: Armed forces in metamorphosis* (pp. 17–34). Berlin: Nomos.

Siebold, G. (2001). Core issues and theory in military sociology. *Journal of Political and Military Sociology, 29*(1), 140–159.

Siebold, G. (2007). The essence of military group cohesion. *Armed Forces and Society, 33*(2), 286–295.

Soeters, J. (2018). *Sociology and military studies: Classical and current foundations.* London: Routledge.

Soeters, J.L. (2008). Ambidextrous military: Coping with contradictions of new security policies. In M. de Boer & J. de Wilde (Eds.), *The viability of human security* (pp. 109–124). Amsterdam: Amsterdam University Press.

Soeters, J., van Fenema, P.C., & Beeres, R. (Eds.). (2010a). *Managing military organizations: Theory and practice.* London: Routledge.

Soeters, J., van Fenema, P., & Beeres, R. (2010b). Introducing military organizations. In J. Soeters, P.C. van Fenema & R. Beeres (Eds.), *Managing military organizations: Theory and practice* (1–13). London: Routledge.

Thornton, P., & Ocasio, W. (2008). Institutional logics. In R. Greenwood, C. Oliver, R. Suddaby & K. Sahlin-Andersson (Eds.), *The Sage handbook of organizational institutionalism* (pp. 99–129). New York: Sage.

Tillberg, P., Tillberg, L.V., Svartheden, J., Hamstrom, B., & Hildebrant, J. (2017). *Mission Afghanistan: Swedish military experiences from a 21st-century war.* Stockholm: Swedish Centre for Studies of Armed Forces and Society.

Tillberg, P., & Tillberg, L.V. (2013). *Mission commander: Swedish experiences of command in the expeditionary era.* Stockholm: Swedish Centre for Studies of Armed Forces and Society.

Tomforde, M. (2009). "My pink uniform shows I am one of them": Sociocultural dimensions of Germany peacekeeping missions. In G. Kummel, G. Caforio & C. Dandeker (Eds.), *Armed forces, soldiers and civil–military relations* (pp. 37–57). Berlin: VS Verlag fur Sozialwissenshafen.

Tresch, T.S. (2007). Multicultural challenges for armed forces in theater. *Military Power Revue der Schweizer Armee, 2*, 34–43.

Urry, J. (2007). *Mobilities.* London: Polity.

Van Dijk, A., Soeters, J.M.M.L., & de Ridder, R. (2010). Smooth translation? A research note on the cooperation between Dutch service personnel and local interpreters in Afghanistan. *Armed Forces and Society, 35*(5), 917–925.

van Fenema, P.C. (2009). Expeditionary military networks and asymmetric warfare. In G. Caforio (Ed.), *Advances in military sociology: Essays in honor of Charles C. Moskos* (pp. 267–283). Bingley, West Yorkshire: Emerald.

de Waal, A. (2013). An emancipatory imperium? Power and principle in the humanitarian international. In D. Fassin & M. Pandolfi (Eds.), *Contemporary states of emergency: The politics of military and humanitarian interventions* (pp. 295–316). London: Zone Books.

de Ward, E., & Kramer, E.H. (2010). Expeditionary operations and modular organization design. In J. Soeters, P.C. van Fenema & R. Beeres (Eds.), *Managing military organizations* (pp. 71–83). London: Routledge.

de Ward, E., & Soeters, J. (2007). How the military can profit from management and organization science. In G. Caforio (Ed.), *Social sciences and the military: An interdisciplinary overview* (181-96). London: Routledge.

2 New directions in military sociology

Reflections on a book project 15 years later

Eric Ouellet

Introduction

In 2005, I co-wrote and edited a book entitled *New Directions in Military Sociology* (Ouellet, 2005). This book brought together several authors who wanted to broaden the horizons of military sociology, from Canada, France, Germany, Russia, Slovenia, the Netherlands, the United Kingdom and the United States. At the time, I sensed that there was indeed a growing but mostly unspoken dissatisfaction with the state of military sociology. I even remember meeting a newcomer to the field at one of the Inter-University Seminar on Armed Forces and Society conferences who told me that he felt military sociology was 20 years behind the rest of sociology. I agreed with him that there was a problem.

It is now 15 years later, and I think it is a good occasion to see if progress has been made, and, if so, how can military sociology be influential where it really counts. This chapter, however, does not propose a review of the current literature in military sociology, but rather a more personal presentation of specific projects in which I was involved, with the purpose of illustrating potential new directions for military sociology. This chapter briefly revisits the more generic problems with military sociology that were highlighted in 2005 and uses them as a template to assess what would constitute a desirable change to the field. Afterward, the chapter will provide a review and assessment of my work on institutional analysis and two NATO projects in which I was involved: Joint Operations 2030 and the Human Environment Analysis Reasoning Tool. Concluding remarks will be on furthering institutional analysis and about the potential offered by the application of design methodologies to military planning, which seems to be one of the leading-edge approaches at the present time.

Military sociology beyond human resources management

In the introductory chapter I wrote for *New Directions in Military Sociology*, I identified some issues that were limiting the scope of military sociology. One of them was the uncritical closeness of many military sociologists to the institution. To be clear, it was not an issue of critical sociology not being interested in the military, as, for instance, there was already research using a feminist approach to study the integration of women in the military at the time. The issue was that a lot of research was sponsored by the military institution, framed and determined based on its perceived corporate needs, essentially in the realm of human resources management. Although there is nothing improper in using sociology for improving personnel management, sociology has a lot more to offer.

This uncritical closeness to the military institution led to an overemphasis on sociological approaches that could be defined as functionalist, in that they tend to remain close to Parsonian sociology, characterized by "the relegation of the specifics of individual relations to a minor role in the overall scheme, epiphenomenal in comparison with enduring structures of normative role prescriptions" (Granovetter, 1985, p. 486). The implications are that, within a human resources management perspective, military personnel are seen as playing a role, and therefore the focus of research is to improve the role-playing adequacy with changes in the corporate environment (such as greater roles for women, minorities, and more educated personnel). These approaches tend to focus on behavioral patterns, taking corporate standards for granted, and unsurprisingly can be quite reductionist in nature. In the context of the Cold War, with a relatively predictable, conventional enemy, such approaches were certainly adequate to meet the needs of the military institution. But, by 2005, a decade and a half after the fall of the Berlin Wall, military sociology had, also, to transform itself for the post-Cold War world.

There is abundant literature discussing the meaning of the post-Cold War world. I will emphasize only two key aspects. The first is that the enemy is now much less predictable, and its face is changing. From the Soviet hordes racing through the Fulda Gap to the Islamic State and its various incarnations, while Russia and China remain potent adversaries, we are in a different world, and the range of military missions has qualitatively enlarged. Keeping an implicit approach based on Parsonian roles does not make much sense. The second aspect, closely linked to the first one, is that those missions are oftentimes not strictly military in nature, involving human security, support for development, humanitarian missions, defense diplomacy, as well as various forms of security-related missions, all of which require close cooperation with civilian counterparts. In such a context, sticking to an implicitly static traditional understanding of military roles tends to be very often counterproductive.

New directions: helping the military to think differently

In 1959, C. Wright Mills published his landmark book *The Sociological Imagination*. This book influenced generations of sociologists and continues to do so. Mills argued that the personal, historical and social all interconnect, and using imagination to find our ways within such contexts was the main task of sociology. In the last year of the second decade of the 21st century, the world has changed, and matters of security have got significantly more complex and challenging; threats against liberal democracy have morphed and became much more diverse. New tasks assigned to the armed forces to defend modern democracies have become daunting. In such a context, it is my contention that military sociology has a key role to play in helping us navigate this world, but both the armed forces and sociology have to become much more imaginative.

Given my position as a professor at Canadian Forces College (Canada's joint staff and war college), one could say that I am quite close to the military institution, and, therefore, I might be unable to be truly critical or imaginative. However, I suggest that, quite to the contrary, I find myself in an ideal personal, historical and social context to be imaginative. I enjoy full academic freedom. My research choices and agenda are not guided by the corporate priorities of the institution. Yet, I am in a position where I know the military institution in much more intimate ways than most sociologists in civilian universities. Hence, I am able to influence it without being constrained by corporate considerations.

This, in turn, has allowed me to focus on what I consider to be one of the greatest challenges for the military in the 21st century: thinking about military operations in ways situated outside classic institutional military roles. In the context of combat formations, this means that commanders and military planners in theater and operational headquarters, because they are facing a wide range of missions and requirements outside classic military roles, have to think and plan differently if they want to set their mission on course for success; however that might be defined.

In reaction to that, some traditionalists may argue that wars are won on the battlefield. This may be true, but they also fail to understand that wars are lost at headquarters owing to poor decision-making and planning, whether it is at the strategic or operational level. Hence, it is my contention that, in the 21st century, military decision-making and planning are the most important "centers of gravity" for combat formations to use classical military terminology.

This is where a renewed military sociology can make a significant impact on combat formations. Since the 1960s, the increasingly preponderant approaches influenced by authors like Berger and Luckmann (1966) and their concept of the social construction of reality, the new sociology and anthropology of knowledge developed by Latour (1987) and Woolgar (1988), and various neo-Kantian authors such as Foucault (1972) and Friedmann (2000), have emphasized that knowledge, perception and understandings of reality are profoundly influenced by social and institutional conditions.

In the context of combat formations, this means reconsidering how we think about issues such as what the mission is (was the Canadian mission in Afghanistan aimed at defeating the insurgency or was it just to be there?); who the enemy is (in Syria, is the Assad regime an enemy in light of all the Islamist forces involved?); what the end state is (does the state of Israel have an end state in its dealings with Hamas and Hezbollah?); what the battlespace is (is the info and cyber war that Russia is waging against Ukraine military in nature, and, if so, what is the military role in it?); and so on. Classic military planning and thinking are very ill-equipped to deal with those modern challenges.

Some directions explored

I was not the only one, in 2005, to identify problems with the thinking in military planning. Some military thinkers were already actively engaged in trying to find ways of planning that are more in touch with the realities of the post-Cold War. Among them, one can highlight the interest in systemic operational design (SOD) popularized by authors like BGen (Retd) Shimon Naveh (1997) of the Israel Defence Force (IDF), and applications of system and complexity theory to military planning (Alberts & Hayes, 2003; Arquilla & Ronfeldt, 2000; Czerwinski, 1998; Moffat, 2003; Smith, 2002). Yet, most of those attempts to rethink approaches to military planning were based on general system theory or engineering-like applications of philosophical or abstract concepts, where the military social roles were still taken for granted, and they essentially focused on the delivery of kinetic force.

However interesting those discussions were at the time, it appears that much of that thinking did not resist the institutional military resilience in not changing. SOD, swarming or effect-based operations are concepts that have lost their attractiveness over the years. The IDF dropped SOD after the war against Hezbollah in 2006, as it found it unsuccessful in predicting the enemy's changes of approach (although it is to be noted that SOD is coming back in favor). Many American concepts were found impractical when its military got deeply involved in counterinsurgency operations in Iraq and Afghanistan. It is my contention that, beyond the normal cycles linked to military intellectual fads, these concepts were doomed as they were divorced from the understanding of social and institutional contexts, and this includes understanding the blue side as well (namely, they were not self-reflexive in that they did not challenge military social roles in a meaningful way).

Institutional analysis

It is in this context that I developed a research agenda using sociological institutional analysis to understand operational failures of combat formations, through a series of historical case studies ranging from colonial times to non-Western peacekeeping operations (Brassard, Dionne & Ouellet, 2019; Chartrand et al., 2017; Conley & Ouellet, 2012; Lacroix-Leclerc & Ouellet 2011; Ouellet, 2008, 2009, 2011, 2013a, 2013b, 2017; Ouellet & Conley, 2013; Ouellet et al., 2013, 2014; Ouellet & Lanteigne, 2011; Ouellet & Pahlavi, 2011, 2015; Pahlavi & Ouellet 2009, 2012). Émile Durkheim famously wrote that sociology is the science of institutions, and it would seem that the sociological study of the military institution would be an obvious application of the discipline. Yet, it is not the case.

It is probably useful to highlight what an institution is. One famous definition is from March and Olsen (2006, p. 3), who wrote that it is a

> "relatively enduring collection of rules and organized practices, embedded in structures of meaning and resources that are relatively invariant in the face of turnover of individuals and relatively resilient to the idiosyncratic preferences and expectations of individuals and changing external circumstances".

Another one is from Durkheim (1895), where he notes that institutions are ways of collecting actions and thoughts (legal, conventional, customary) that have their own existence, outside individuals, and tend to be spontaneously conformed to; they exert on individual consciousness a coercive influence, with beliefs and social practices acting from outside in.

These definitions, and most other definitions of institutions, point to a few key issues that do not sit well with military ways of thinking and, hence, might explain the limited use of sociological analysis over the years to understand the military institution. First, it implies that there are forces external to individuals shaping their ways of thinking and their behaviors. This goes against one of the central myths in military affairs: the commander (and his or her staff) as a rational and free agent who determines military courses of action. Worse, those forces act in ways that are both subtle (if not unconscious) and coercive (going against them leads to possible punishment), and they are enduring and hard to change. Hence, decisions must be taken at times to avoid creating institutional problems rather than finding a solution to the problem at hand: to deal with multiple, and at times contradictory, institutional logics. Many aspects of decision-making are therefore outside the realm of the rational. Commanders are not free agents, and they are bound to make non-rational decisions. This explains in part why classic military social roles are so pervasive, as they may provide a false sense of stability and comfort in the face of conflicting institutional dynamics.

Yet, institutional analysis, especially in its more recent versions, allows for the possibility to make social agents aware of their own role, therefore potentially regaining some degree of agency, even within a strong institution like the military. In other words, institutional analysis can help commanders and planners of combat formations to determine courses of action in ways that are much more self-reflexive and open to notions from outside their institutional roles.

The institutional analysis approach I use is very much influenced by the model developed by Richard Scott (2008). It uses a three-dimensional approach. The first is defined as the regulative and emphasizes decisions that are justified based on both formal and informal rules, regulations, laws and sanction systems. The second one focuses on normative aspects, looking for implicit values and norms that are collectively shared about what is desirable, acceptable and "right". Implicitly favoring conventional military approaches over irregular ones in decision-making would be a good illustration of normative institutional forces in action. Decision-makers are oftentimes not fully conscious of these aspects, which can be assessed indirectly by interpretative approaches. The last dimension is the cultural-cognitive one and refers to shared preconceived notions, thought patterns and worldviews that are invoked in decision-making. Invoking past military practices to justify a course of action is a classic example of such. Combined together, these three dimensions of institutional analysis provide a comprehensive framework for understanding the actions and decisions of combat formations. The central unit of analysis in the institutional analysis consists of the key decisions taken by the major actors, which lead to real action or inaction.

To highlight this approach in more concrete terms, here are two examples drawn from my research in decision-making by combat formations, using institutional analysis. The first one is the case of the IPKF (Indian Peace-Keeping Force) in Sri Lanka from 1987 to 1990 (Ouellet, 2011). What was at first construed as a relatively benign peacekeeping mission turned into a full-fledged counterinsurgency operation that has been qualified by many as India's Vietnam. What is more troubling is that the Indian Army has had a long history of relatively successful counterinsurgency operations domestically. Yet, when it faced the Tamil Tigers of the LTTE (Liberation Tigers of Tamil Eelam) it failed badly. A number of thoughtful analyses were proposed to explain such an outcome, but they did not put much emphasis on the non-military challenges specific to counterinsurgency.

The IPKF had to switch very quickly into a combat role, as the LTTE decided to fight the Indian forces. Following a costly attempt by the IPKF to take control of Jaffna in October 1987, the LTTE retreated to the inland jungles. For the next three years, the IPKF had to deal with an insurgency that it failed to overcome, and, in the end, it had to leave Sri Lanka in shame. The period of the post-Jaffna battle was marked by repeated cordon search operations, oftentimes conducted at the scale of brigade- or division-size forces. Given that large formations were used for those counterinsurgency operations, it was relatively easy for the LTTE to anticipate the Indian Army's movement and escape. Once over, the LTTE could resume its hit-and-run harassment operations against the IPKF. A number of Indian officers were aware that these actions were ineffective and called them "jungle bashing". Yet, the Indian combat formations continued to pursue the same approach and planning, in spite of their obvious failure.

Through institutional analysis, it was possible to see that IPKF commanders and planning staff had to contend with two mutually opposing institutional forces, and jungle bashing became the strange compromise to keep those forces in balance. From a normative standpoint, the Indian Army was struggling to link internal and external legitimacy. On the one hand, the institution was trying to establish a legitimate external normative image for itself (i.e., engaged in a peacekeeping mission instead of waging a war on foreign soil). On the other hand, it was a proven conventional army unable to defeat a small but clever and innovative enemy, and thus was facing an internal legitimacy challenge to its own military competency. The extensive use of jungle bashing could be understood as a "compromise" between an internally driven response to the environment (giving the appearances of a conventional army doing "army things") and a normatively driven need to maintain external legitimacy (focusing on the "enemy of peace", not the local population, i.e., closer to a classic peacekeeping role), even if such "compromise" was profoundly counterproductive.

As the IPKF was engaged in a counterinsurgency operation outside India, its cultural-cognitive dispositions toward social isolation and apolitical outlook were also highly problematic. As has been written about extensively on counterinsurgency operations, these are profoundly political in nature because they involve getting the local population on the side of the governmental forces in order to get actionable human intelligence. If seeking out population support could be done by the Indian Army in a domestic context, to do that in a foreign land was something quite different. Given that the governments of India and Sri Lanka, and the LTTE, were all playing a political game of their own, the cognitive "reflex" of being the apolitical peacekeeping force was a recipe for disaster for the Indian Army, as any of its actions could be undermined politically by everyone else. Yet, to maintain both its internal and external legitimacy, the IPKF could not engage the conflict from a political perspective either; it had to remain being seen as apolitical. Hence, the blind focus on jungle bashing was also justified by cognitive institutional dynamics, as it could justify its role as a purely military and non-political one.

The second example is drawn from an institutional analysis of the IDF operations against Hezbollah in 2006 (Pahlavi & Ouellet, 2012). In this example, the challenge appears that IDF combat formations involved in the Lebanon War of 2006 were over-adapted for "smart" counterterrorism operations, leaving commanders and planning staff unable to use their forces in ways that would have been more effective. On 12 July 2006, Hezbollah fighters crossed the Israeli–Lebanon border and launched a daring raid on an Israeli patrol. In the clash that ensued, three Israeli soldiers were captured, and eight were killed. This provocation acted as the spark embroiling the two sides, Israel and Hezbollah, in a bloody war that would later become known as the 33-Day War, or the Second Lebanon War.

Resulting in nearly 1500 fatalities on both sides, the conflict itself could be described as two distinct segments. The first one was led by the Israeli Air Force (IAF), which launched a wide air strike campaign destined to destroy the physical infrastructure of Hezbollah in Beirut and southern Lebanon, as well as the communications infrastructure, roads and bridges linking southern and northern Lebanon, to physically isolate Hezbollah. Although the air campaign caused significant physical damage in Lebanon, it did not prevent Hezbollah from continuing its activities. The second segment was essentially led by the army through a mechanized advance in southern Lebanon to destroy Hezbollah on the ground and its rocket-launching systems, which were being used against civilian areas in Israel. The ground campaign was also problematic, as Hezbollah surprised the Israelis by sending out aggressive tank-hunting teams to stop the mechanized advance and established well-designed fortifications around its strongholds. Neither the IAF nor the army achieved their objectives. Given very strong international pressure on Israel to stop, and fearing a costly, protracted war, the Israeli government decided to put an end to the operations after 33 days. The war ended with a weaker Israeli deterrence and greater Hezbollah prestige in the region.

How can we make sense of the difficulties the IDF faced during the 33-Day War, despite its long experience with various irregular conflicts? To answer this question, it is necessary to go beyond classical explanations limited to logistical, doctrinal and geopolitical factors and shed light on the interaction of institutional forces, which are, in general, neglected in the analysis of military affairs, and how they can have a very direct impact on combat formations. The Israeli society at large and the IDF as an organization had both undergone tremendous transformations in the two prior decades, which had significant effects on both the IDF combat formations and their potential for adaptation. Foremost among these changes has been the transformation of Israeli society from a "nation-in-arms" to a "post-heroic" model increasingly sensitive to casualties. More liberal, individualistic-oriented values emerged. The effects of such social changes on the IDF are inextricably linked to the technocentric approaches espoused by the Israeli military.

"High tech" warfare is oftentimes perceived as "touch less" warfare, where casualties are minimized and a greater distance between the troops and the actual violence of war is established. Hence, it is no surprise that technocentrism had a deep impact on the IDF's doctrinal and organizational dimensions. Material constraints coupled with the consolidation of the post-heroic value systems reinforced normative institutional forces, pushing the IDF to undergo a transformation leading to the establishment of a "smaller but smarter" IDF. Furthermore, on the regulative level, the unique nature of civil–military relations in Israel, where politicians are oftentimes former senior military commanders, further compounded such institutional forces by not providing enough of the necessary challenge functions to military decision-makers. Lastly, on the cognitive level, the environment did not provide enough pressure on the IDF to force any institutional adaptation, as the handling of the Second Intifada was deemed successful and, as such, prevented the IDF from contemplating that the next war might be quite different from the last. The IAF and the Army engaged in a "high tech" conflict because, institutionally, it was the right compromise for the IDF at the time. The combat formations were stuck in a military role that aimed at stopping a violent social movement, instead of seeking to achieve effects that would undermine its enemy politically; so too were the decision-making and planning.

In many ways, sociological institutional analysis can act as a form of "organizational psychoanalysis", as was suggested many years ago (Lapassade, 1967; Lourau, 1970). This approach can help the military to come out of denial about its understanding of reality, its practices, its blind spots, and so on. At the operational level, it is certainly an approach that can help combat formation commanders and planning staff "to think outside the box". Given that the world seems unwilling to stop changing, I sense that there is more willingness on the part of the "patient" to explore its own denials than in the past. Certainly, most of my military students at the Canadian Forces College see substantive value in such an approach, once it is mastered.

NATO projects

I have also ventured in new directions in military sociology through two NATO-sponsored research projects. The projects I was associated with shared the same goal and were initiated for the same reasons: to broaden the horizons of military planners by the injection of social sciences, and to develop planning processes better informed by social sciences.

The first project I was involved with was sponsored by the NATO Research and Technology Organization (RTO; now known as the Science and Technology Organization – STO) and was entitled Joint Operations 2030 (NATO RTO, 2011a). This project was aimed at identifying future military capabilities that will be required by 2030, and it was born out of a request from the NATO National Armaments Directors to influence the future of military procurement. The project was initiated in 2006 and involved the participation of experts from 13 NATO countries and from five NATO agencies. The group was made up of military officers and natural and social scientists. My own involvement was mostly in the upfront part, to ensure that the project got off to a good start, as Canada had the formal lead on the project. The first challenge was to ensure that the group did not mentally become a prisoner of traditional military roles. After some key presentations, several meetings and numerous informal exchanges, this first hurdle was overcome, and this led the group to develop some innovative concepts. In particular, they developed the notion of "thematic analysis" to look at the potential combat environment of the future.

An important epistemological challenge of this project was about how to assess what the future might be, an extremely difficult task. The group concluded that linear analyses like trend analysis and technology watch were not satisfactory and went against the most recent findings in futurology. It is in this context that the concept of theme emerged. Themes were selected based on implicit criteria that some present-day or emerging challenges are likely to remain potent for the long term. The real, unknown issue was how they might interface with future social and political dynamics. Hence, the themes were implicitly constructed around the generic capacity of military and security institutions to react and adapt to those challenges. Any inadequacies between persistent or emerging challenges and social and institutional setup would then, in turn, inform experts on what would be required for the military of the future.

Although this project was more geared toward supporting defense planners at the strategic level, rather than military planners in combat formations, it produced some interesting sociologically informed concepts and notions that can be useful to combat formations. As stated in the report,

> Employing a series of creative spirals and iterations the JO 2030 Study agreed upon a set of 18 Themes. These Themes do not define the future, are not intended to be comprehensive in their scope or coverage. They were meant to be provocative and to be distinct from well-known and studied trends that many experts have clearly agreed will have an impact on future operations such as nanotechnology, climate change, or robotics to name but a few. In summary the Thematic Analysis was a deliberate effort to not simply validate the well-known problems of tomorrow but rather to raise the level of discussion of the not so well-known challenges that may develop or evolve with time.
>
> (NATO RTO, 2011a, pp. 4–1).

Those themes ranged from non-military/non-violent threats to national staying power to planning in the face of deep uncertainty. They were broad, yet quite realistic in nature, and involved having a much more honest look at the challenges that NATO might face in the future, including the inherent limitations of NATO's member states. For instance, Theme 17, The Role of Information and the Media – quite prescient of the Russian operations in Ukraine five years later – was described as follows:

> The media has become instrumental in developing the context for the public audiences that affect the Alliance. The pervasive 24/7 media cycle will continue to create the "CNN effect" where strong emotional content can engender public reaction which may affect political and military decision-making at all levels of command. There is a symbiotic relationship between the military and the media in that the media requires access and information and the military needs the media to communicate with the public. The increased instantaneous access to information available to the public will be a serious consideration in the future as public perception can drive constraints on both the political and military levels.
>
> (NATO RTO, 2011a, pp. 4–5)

The key strength of the thematic analysis methodology was that it forced military planners to take a multidisciplinary approach, which led to integrating sociological and political considerations into planning, in ways that are holistic in nature, and steering planners away from strictly kinetically based military roles.

As they integrate an increasing number of civilian specialists into their headquarters (political advisors, gender advisors, cultural advisors, as well as reaching out to experts at their respective MoD), combat formation commanders and military planning staff could replicate thematic analysis at their level, especially in the context of the operational level of war. For instance, during the 2011 operations in Libya, it would have been useful to develop thematic analyses for the NATO operational commanders on a "post-operation Islamist Libya", a danger that many experts were warning NATO about during the operations. This might have changed how NATO supported anti-Gaddafi forces, right down to the tactical level.

It is unclear how much take-up of thematic analysis followed the production of the report. The 2011 Libyan operations were a watershed for NATO, which recognized that it was ill-prepared to deal with this kind of mission. During a research trip to NATO's Supreme Headquarters of Allied Powers in Europe in 2014, I was briefed on the emergent Comprehensive Crisis and Operations Management Centre (CCOMC), which was created to deal with operations like the one in Libya, using a cadre of social scientists to keep a tab on potentially emerging crises in the world. CCOMC would have been an excellent venue to test thematic analysis and provide NATO's combat formations with wider perspectives on upcoming challenges. Although such a center can be useful, it remains focused on short-term, potential applications of kinetic energy.

The second NATO project I was involved in was the Human Environment Analysis Reasoning Tool (HEART; NATO RTO, 2011b). As stated in the report, the tool was designed to help

> military staff and analysts understand the human and social environment, develop effective courses of action, and make use of appropriate analysis methods. HEART can be used to support early phases of operational planning, as a training resource and to improve the representation of the human and social environment in exercises and experiments.
>
> (NATO RTO, 2011b, p. ES-1)

This project was developed with combat formation commanders and planning staff in mind. The project started in 2008 and was very much influenced by the events of that time. The difficulties of NATO in Afghanistan and the American-led coalition in Iraq, dealing with the social, cultural and political aspects of military operations, created greater demands for sociological involvement in military affairs. Although the United States had deployed several human terrain system teams in both Afghanistan and Iraq, their successes were mostly localized at the tactical level. Operational-level planning and decision-making still seemed very disconnected from a sound understanding of non-kinetic matters. To put it another way, traditional military institutional roles at the operational level of war were much harder to overcome.

I was involved in the entire length of this project, from the pre-project definition phase to the production of the final report. The project was led by the United Kingdom but had a substantive American involvement and included experts from Germany, the Netherlands, Norway, Sweden, and, of course, Canada. The early days of the project reproduced, at its level, some of the challenges of dealing with functionalist social sciences and their taken-for-granted assumptions. Experts came from different disciplines and schools of thought, which threw the project into a deep epistemological and ontological debate during the first year, and it almost derailed everything. A consensus was eventually reached that the tool would have to be flexible enough to allow for different approaches and perspectives to be available to military planners, but in the end the goal remained the same: military planners in combat formations had to think differently about their military and social role while planning.

The group was also pragmatic and recognized that institutional roles could not be overcome by this project alone, and a degree of the institutional role had to be embedded within the tool. Hence, the tool was purposefully designed using the NATO formal operational planning process (OPP) to ensure that the narrative surrounding the tool would be cognitively perceived as institutionally legitimate by potential users. The tool was particularly designed to help with the concept development phase of the OPP. Furthermore, given the scope of the project, only a proof of concept could be developed by the group. The tool was developed using a visual aid software called CMap, where useful sociological, anthropological, psychological, and so on, concepts, notions, guiding questions, examples and links to further online resources were embedded in logically cascading pages.

As the name of the tool implies, it was an aid to reason, to think about aspects, dimensions and dynamics which are normally ignored or only partially covered through the normal military planning process. If the group had had more resources, then it could have made it more user-friendly and filled with easy to use but sociologically meaningful templates and checklists for military planners.

The main challenge with the tool is that there is a lot of implicit knowledge built in, even in the simplest sociological description, and it would be quite unrealistic to expect military planners to be experts in sociology or anthropology. Finding the relevant sociological data to answer questions raised by the tool, knowing where to find it and how to digest it quickly without extensive expertise in social science would be problematic. Hence, the tool was perceived as something that could initiate thinking from a different perspective, outside the institutional role, and hopefully provoke enough interest that military planners would reach out for expert help as needed. But it was not aimed at planners in tactical headquarters engaged in fast planning cycles.

The tool was positively received by the Allied Command Transformation in Norfolk, Virginia. To go beyond the proof of concept, however, would require a significant investment in money and staff time. Given that institutional social roles are quite resistant in the military, the next step for a HEART-like tool will probably require another major conflict to occur, where sociological knowledge is once again badly needed and yet still in short supply.

Both projects, like most NATO RTO-sponsored projects, have results that are short-lived, not so much because of a lack of quality, but because of the institutional logic of the member states' defense research establishment to roll out project after project, to justify their usefulness. Yet, in my opinion, the true usefulness of these projects lies in spreading ideas and notions through their international networks of experts, who in their turn can influence their own respective armed forces to think about problems and issues differently.

Future directions in military sociology

Ill-designed design?

One of the emerging approaches in the military world to improve combat formation commanders' decision-making and staff planning is known by the notion of "design methodology" (Banach & Ryan, 2009) or simply "design" (Perez, 2011). The present-day interest in design applied to military decision-making and planning is often perceived as a continuation of the original ideas proposed by Naveh on systemic operational design more than a decade ago (Beaulieu-Brossard & Dufort, 2017). It has become an official approach with the US Army (2015) and was formally explored by the US Special Operations Forces Command in 2016 and formally embraced in 2018 (Black et al., 2018). In 2017, the *Journal of Strategic and Military Studies* published a special issue on the "reflexive turn" in military affairs, where design is the anchoring concept of most of its articles. The

Australian military has recently published a book on the use of design for military affairs (Australia, 2019). It seems on the surface that the message of improving military thinking has been heard. Yet, it also appears that military sociology still has its work cut out.

The concept of design as an approach to develop innovative solutions to practical problems has been effectively used for quite some time, and it is well established in numerous fields, such as architecture, marketing and interior design and is becoming increasingly adopted in other fields, such as in information systems. There is no consensual definition of design; hence, it might be more challenging to assess it in a military context. However, one can note that design usually implies that something is done, always having in mind the desires and preferences of the end users. The Viennese architect Alfred Loos is famous for having captured the spirit of design more than 100 years ago in saying that "the house did not belong to art because the house must please everyone, unlike a work of art, which does not need to please anyone". Hence, the design is not seeking an ultimate epistemological (or esthetical) solution, but rather the practical arrangements that will serve and please end users. It is, in spirit, a pragmatic approach. As noted above, marketing has embraced design for many decades now, as it carries out surveys and usually tests products with a sample of potential users prior to launch, making corrections and modifications as necessary.

As a creative and healthy move, some military thinkers, mostly in Israel and the United States, started to import some of the ideas and concepts of design to improve military planning and decision-making. This has evolved over time, and design has been further adapted to military affairs, but it has also been controversial (Proctor, 2011). As well, the use of the design in military affairs varies greatly from one service to the next, and from military to military. It is, at present, very much army-centric. Furthermore, despite being more sociologically informed (see, for instance, Paparone (2012)), and often referring to reflexivity in quoting Donald Schön's (1983) well-known interpretivist approach to organization studies, the way military organizations adapt design to military affairs too often still takes for granted classic institutional military roles that are much less adapted to 21st-century missions. A good example of this attitude was seen in a seminal article written by Grigsby and others (based at the famous US Army School of Advanced Military Studies), which was the departure point for "regimenting back" design into a formalized planning process (Grigsby et al., 2011), which offers very few opportunities to challenge institutional roles.

It appears now that there is a desire to "de-regiment" design. The ever-growing military literature on design is fascinating, and it still shows a fair degree of internal debates about how design can be useful to military planning and decision-making. As well, the actual practice of design as an approach in the military lags considerably behind the intellectual debate. However, one common but rarely discussed issue behind this debate is that, too often, it still implicitly considers the end user as being the military itself, usurping the very essence of design by emphasizing the "how" over the "for whom".

For instance, it was noted that:

> Unlike detailed planning, design as an emerging practice evokes eclectic combinations of philosophy, social sciences, complexity theory, and often improvised, unscripted approaches in a tailored or "one of a kind" practice. [...] Institutionally and as a practicing community of professionals, the Western military has little trouble agreeing upon the general principles of traditional planning. Yet we collectively remain fiercely divided, confused, and often resistant to design in any form, whether a rival methodology, complimentary, or even a subset of traditional planning.
>
> (Zweibelson, 2015, pp. 12–13)

In all fairness, some end users such as the political decision-makers are not necessarily able or even willing to express more clearly what they want, and therefore the designer must use a more empathetic approach to figure out what the end users want. This, in turn, requires designers to be more self-reflexive, with the purpose of separating the designer's own preferences from those of the end users that can be detected. As noted above, the military literature on design also emphasizes reflexivity, yet it is too often construed as a matter of individual psychology and self-cognition (Banach & Ryan, 2009), instead of emphasizing the gain in reflexivity through direct contact with others (the actual end users) while bracketing the designer's own preferences and views. As well, military commanders and planners are rarely offered ways of challenging, in an uncomplacent way, the shortcomings of the subtler yet very powerful institutionally based views, perspectives, values and norms that shape decision-making and planning. Institutions do not exist in a vacuum; their methods exist to justify themselves in the eyes of others outside the institution. Sociologically informed reflexivity, especially in the context of going outside classic institutional military roles, is too often lacking.

Another example, drawn from a recent article, implicitly highlights this problem; the author noted that:

> Design should be a liberating experience. Systemic inquiry should be a liberating experience. However, it comes with a price. Liberation is War, surfacing confrontation that most institutions cannot stomach. Not all people want to be liberated. Some rather take the blue pill and remain prisoners of their own mind (or doctrine). Those who do, condemn themselves to a life of frustration, hunger, and discontent. [...] That is why Systemic Design Inquiry is measured by the degrees of freedom it creates! SDI aims at getting our designers on the path of self liberation – far beyond what they know, beyond their experience, value systems, beliefs, prejudices. Beyond doctrine. In order to achieve desired degrees of freedom one must first identify his/her biases, prejudices and axioms carved in institutional "stone". These are the borders they must transgress in order to be liberated.
>
> (Graicer, 2017, p. 35)

Although this excerpt from the military design literature highlights a desire to rise above social and institutional military roles, it remains based on altering roles by taking an inner look at some institutional conditions, instead of looking outward (namely at what legitimizes those institutional conditions). If there is something that sociology has taught us, it is that social roles are defined by our relationship to others, not to ourselves. Hence, even today's leading-edge thinkers of military design still struggle with the fundamental issue of using design that implicitly has the military itself as the end user. This, in turn, raises more questions about why there is this persistent blind spot, or some would say denial, about not providing actual room to alter social, military roles.

Lastly, this denial can also be indirectly observed in many debates found in that intellectual movement, as they center around sophisticated epistemological debates and argumentation, sometimes quite abstract in nature, drawn from philosophy (Clermont, 2017) and system theory, such as how to deal with non-linear problems (Zweibelson, 2016). Those debates are too often disconnected from a sociological institutional understanding of military reality and, hence, keep the classic military social roles protected (or in denial), as they remain undiscussed.

It is in this intellectual atmosphere that military sociology should continue to emphasize helping commanders and staff of combat formations to think in ways that are truly sociological in nature. I remain firm in the conclusion that institutional analysis, by being uncomplacent about institutional roles, is an approach which has a lot to offer in this regard and can provide much-needed food for thought to military designers. As well, institutional analysis can be quite resilient to intellectual fads.

Institutional analysis 2.0

The work done so far using institutional analysis has barely scratched the surface of that potential. A lot more work in fleshing out institutional sociological analysis applied to combat formations, and military affairs in general, remain to be done. Among the future directions in military sociology I am taking are greater incorporation and adaptation of concepts from organization studies. Military commanders and their staff are facing what Friedland and Alford (1991) have described as "institutional logics" or rule-like patterns of thinking that are legitimizing within the institution but are not necessarily adapted to deal with challenges emanating from the environment (like the enemy).

In the same vein, some authors (Greenwood et al., 2010, 2011) have looked into "institutional complexity" as a way to describe very common situations where multiple institutional logics work at cross-purposes (such as obtaining military success versus achieving political/diplomatic goals). Other authors (Thelen, 1999, 2003) have emphasized that institutional logics are not only multiple and create complexity, but they also have various degrees of visibility and ways of affecting decision-making and planning. These logics can be construed as "piling up" over time, or more precisely creating what was termed "institutional layering". In a military context, the older normative institutional logics inherited from the 18th and

19th centuries of quasi-esthetical regularity in forms and structures continue to play an implicit role in determining command and control arrangements between tactical units and their operational headquarters, whereas the more recent logic of being "lean and mean" remains the formal expectations to constitute task forces.

As well, the multiplicity of institutional logics that are working both in parallel and in layered ways, creating complexity, also creates room for individual actors to be creative, either for themselves, or for the group they are identified with, and even for the whole institution. This has been termed "institutional work" (Lawrence & Suddaby, 2006) and emphasizes the capacity of agents to use the multiple institutional rules, narratives, norms and values against each other to act and introduce necessary changes. This is an important dimension of institutional analysis, as it adds a clear set of criteria for an agency. This approach has been further developed under the name of "embedded agency" (Delbridge & Edwards, 2013) and provides a fair bit of potential for understanding the work and social roles of combat formations commanders and planning staff, as they have to be creative in developing "institutional compromises" between various logics.

These various concepts, however, have been developed mostly in the context of the firm, a component of the market as an institution, and to a lesser extent in the context of public administration, part of the institution of the state. Although the military is influenced to some degree by the market, and much more by the state, it remains a distinct institution with its own institutional logics. In fact, if it does not like the state, it can attempt to change it, as seen recently in Turkey, and can support the state in changing the nature of the market, as seen in Venezuela in 2020. The key here is that a careful adaptation of those concepts for the military institution will have to be put in place.

The last direction that I am exploring is more speculative and probably riskier, but at the same time it also appears to be more necessary. It is the role of institutional blind spots or collective denials. As noted above, authors such as Lapassade (1967) and Lourau (1970) have developed a psychoanalytical approach to organizations, which is also called "socio-analysis". Such an approach, if modernized by removing the excessive reliance on Freud, and the Marxist overtone and mythology, could be of use in helping to identify and bring back to consciousness deeper normative institutional logics that remain mostly unconscious among combat formations commanders and planning staff.

In the same vein, some of the concepts and methodologies developed by people in group analytics about the social unconscious, following the landmark work of S.H. Foulkes (1964) and his followers (Hopper, 2003), can be a quite useful complement to socio-analysis. Foulkes developed group psychoanalysis during World War II in order to be able to treat larger numbers of soldiers with psychological injuries, as the need vastly exceeded the capacity to offer individual psychoanalysis as standardized by Freud. With time, this approach was used for people having common treatment needs such as alcoholics, victims of sexual abuse, and so on. What is central in the group analytics is that social pressures to conform to certain dysfunctional standards are the key obstacles to treatment. These pressures are internalized unconsciously but are shared socially by most members of society.

Hence, by using group methodology, the analyst can help the group to see their common challenges, which they are all in denial about. Taken away from their clinical purposes, these concepts and methodologies can be useful in helping to map, identify and raise to shared consciousness unconscious military roles and unconsciously accepted institutional logics.

Conclusion

Fifteen years after the publication of *New Directions in Military Sociology*, it seems that certain things have changed, whereas others are not that much. I observed a much wider variety of sociological approaches in journals such as *Armed Forces and Society*, and also in presentations at the Inter-University Seminar on Armed Forces and Society conference. Some of the "lessons learned" by the military in the wake of the Afghanistan, Iraq, Libya and Syria missions about the importance of the social and cultural context in operations and recognition of the inherent operational differences involved in non-conventional warfare seemed to have resisted the test of time.

On the other hand, the need to help military commanders and planning staff to think differently, given the expanded scope of missions they are given, remains an important preoccupation without much resolution. As in the past, it seems that this concern is mostly shared by current or retired military officers and civilian professors in military colleges, most of them emphasizing approaches from either philosophy, system theory and cybernetics, cognitive psychology or strategic studies. Military sociologists venturing to consider those questions are, still, few and far apart, even if their contribution could be quite substantive.

In the end, military institutional roles are still rarely discussed in sociologically imaginative and uncomplacent ways. To do so means unveiling some implicit forms of legitimation for the military institution, which in turn can be construed as risky and destabilizing by its members. Yet, the military institution, like any other institution, is called on to change because its environment is changing. Institutional change can be both abrupt after a major crisis, and incremental and on an ongoing basis. Military sociology can help in making such changes less painful.

Sociological institutional analysis applied to the military appears to still have a nearly untouched field for future research. Combat formations are likely to continue morphing into new roles and missions, thus bringing new challenges. Yet, unless the military institution ceases to exist as we understand it, institutional adaptations and strange compromises will continue to be defining features of military affairs in the 21st century. This is why institutional analysis, as one of the new directions in military sociology, appears to me to be a pretty safe bet, and a very much-needed tool.

References

Alberts, D.S., & Hayes, R.E. (2003). *Power to the edge*. Washington, DC: Command and Control Research Program.

Arquilla, J., & Ronfeldt, D. (2000). *Swarming and the future of conflict*. Santa Monica, LA: RAND.

Australia. (2019). *Design thinking: Applications for the Australian Defence Force* (Joint Studies Paper Series, no. 3). Canberra: Australian Defence Publishing Service.

Banach, S.J., & Ryan, A. (2009, March–April). The art of design: A design methodology. *Military Review, 87*, 105–115.

Beaulieu-Brossard, P., & Dufort, P. (2017). Introduction: Revolution in military epistemology. *Journal of Strategic and Military Studies, 17*(4), 1–20.

Berger, P.L., & Luckmann, T. (1966). *The social construction of reality*. London: Penguin.

Black, C., Newton, R.D., Nobles, M.A., & Ellis, D.C. (2018). U.S. special operations command's future, by design. *Joint Force Quarterly, 90*(3), 42–49.

Brassard, C., Dionne, C., & Ouellet, E. (2019). La France et l'OTAN : Une continuité inopinée. In *NATO at 70 years: Selected topics in world security* (pp. 9–12). Toronto: NATO Association of Canada.

Chartrand, P., Harvey, F., Tremblay, E., & Ouellet, E. (2017). North Korea: Perfect harmony between totalitarianism and nuclear capability. *Canadian Military Journal, 17*(3), 29–39.

Clermont, F. (2017). Design: An ethical and moral project conscious intention for the cybernetician. *Journal of Strategic and Military Studies, 17*(4), 51–83.

Conley, D., & Ouellet, E. (2012). The Canadian forces and military transformation: An elusive quest for efficiency. *Canadian Army Journal, 14*(1), 71–83.

Czerwinski, T.J. (1998). *Coping with the bounds – Speculations on nonlinearity in military affairs*. Washington, DC: Institute for National Strategic Studies.

Delbridge, R., & Edwards, T. (2013). Inhabiting institutions: Critical realist refinements to understanding institutional complexity and change. *Organization Studies, 34*(7), 927–947.

Durkheim, É. (1895). *Règles de la méthode sociologique*. Paris: Alcan.

Foucault, M. (1972). *Histoire de la folie à l'âge classique*. Paris: Gallimard.

Foulkes, S.H. (1964). *Therapeutic group analysis*. London: Allen and Irwin.

Friedland, R., & Alford, R.R. (1991). Bringing society back in: Symbols, practices, and institutional contradictions. In W.W. Powell & P.J. DiMaggio (Eds.), *The new institutionalism in organizational analysis* (pp. 311–336). Chicago, IL: University of Chicago Press.

Friedmann, M. (2000). *A parting of the ways: Carnap, Cassirer, and Heidegger*. Chicago, IL: Open Court Publishing Company.

Graicer, O. (2017). Self disruption: Seizing the high ground of systemic operational design (SOD). *Journal of Strategic and Military Studies, 17*(4), 21–37.

Granovetter, M. (1985). Economic action and social structure: The problem of embeddedness. *American Journal of Sociology, 91*(3), 481–510.

Greenwood, R., Díaz, A.M., Li, S.X., & Lorente, J.C. (2010). The multiplicity of institutional logics and the heterogeneity of organizational responses. *Organization Science, 21*(2), 521–539.

Greenwood, R., Raynard, M., Kodeih, F., Micelotta, E.R., & Lounsbury, M. (2011). Institutional complexity and organizational responses. *Academy of Management Annals, 5*(1), 317–371.

Grigsby, W.W., Gorman, S., Marr, J., McLamb, J., Stewart, M., & Schifferle, P. . (2011, January–February). Integrated planning: The operations process, design and the military decision-making process. *Military Review, 89*, 28–35.

Hopper, E. (2003). *The social unconscious: Selected papers*. London: Jessica Kingsley.

Lacroix-Leclair, J., & Ouellet, E. (2011). La Petite guerre' en Nouvelle-France 1660–1759: Une analyse institutionnelle. *Canadian Military Journal, 11*(4), 48–54.

Lapassade, G. (1967). *Groupes, organisations, et institutions.* Paris: Gauthier-Villars.

Latour, B. (1987). *Science in action.* Cambridge, MA: Harvard University Press.

Lawrence, T.B., & Suddaby, R. (2006). Institutions and institutional work. In S. Clegg, Hardy, C., Lawrence, T., & Nord, W.R. (Eds.), *Handbook of organization studies* (pp. 215–254). London: Sage.

Lourau, R. (1970). *L'analyse institutionnelle.* Paris: Editions de Minuit.

March, J.G., & Olsen, J.P. (2006). Elaborating the "new institutionalism". In R.A.W. Rhodes, S.A. Binder & B.A. Rockman (Eds.), *The Oxford handbook of political institutions* (pp. 3–20). Oxford: Oxford University Press.

Moffat, J. (2003). *Complexity theory and network centric warfare.* Washington, DC: Library of Congress – DoD Command and Control Research Program.

Murden, S. (2013, May–June). Purpose in mission design: Understanding the four kinds of operational approach. *Military Review, 91,* 53–62.

Naveh, S. (1997). *In pursuit of military excellence: The evolution of operational theory.* London: Frank Cass.

NATO RTO. (2011a). *Joint operations 2030 – Final Report.* Paris: North Atlantic Treaty Organization - Research and Technology Organization. Report TR-SAS-066.

NATO RTO. (2011b). *The human environment analysis reasoning tool (HEART) – Incorporating human and social sciences into NATO operational planning and analysis* (Report TR-SAS-074). Paris: North Atlantic Treaty Organization - Research and Technology Organization.

Ouellet, E. (Ed.). (2005). *New directions in military sociology.* Toronto: de Sitter.

Ouellet, E. (2008). Ambushes, IEDs and COIN: The French experience. *Canadian Army Journal, 11*(1), 7–24.

Ouellet, E. (2009). Multinational counterinsurgency: The Western intervention in the Boxer Rebellion 1900–1901. *Small Wars and Insurgencies, 20*(3/4), 507–527.

Ouellet, E. (2011). Institutional analysis of counterinsurgency: The case of the IPKF in Sri Lanka (1987–1990). *Defence Studies, 11*(3), 470–496.

Ouellet, E. (2013a). Les forces armées comme institution. In E. Ouellet, P. Pahlavi & M. Chenouffi (Eds.), *Études stratégiques au 21ème siècle* (pp. 263–285). Montréal: Athéna.

Ouellet, E. (2013b). Les années 1990: Émergence du soldat-diplomate. *Guerres Mondiales et Conflits Contemporains, 250*(2), 55–65.

Ouellet, E. (2017). The self and the mirror: Institutional tensions and the Canadian Special Operations Forces. In J.G. Turnley, K. Michael & E. Ben-Ari (Eds.), *Social sciences and Special Operations Forces* (pp. 185–200).

Ouellet, E., & Conley, D. (2013). Les Forces canadiennes et les défis institutionnels des années 1960–1970. *Guerres Mondiales et Conflits Contemporains, 250*(2), 41–54.

Ouellet, E., Lacroix-Leclerc, J., & Pahlavi, P. (2013). Institutionalization of irregular warfare: The case of Darfur. *Contemporary Military Challenges, 15*(3), 11–24.

Ouellet, E., Lacroix-Leclerc, J., & Pahlavi, P. (2014). The institutionalization of al-Qaeda in the Islamic Maghreb. *Terrorism and Political Violence, 26*(4), 650–665.

Ouellet, E., & Lanteigne, P.M. (2011). Institution militaire et contre-insurrection: L'IPKF (1987–1990). *Guerres Mondiales et Conflits Contemporains, 241,* 107–123.

Ouellet, E., & Pahlavi, P. (2011). Institutional analysis and irregular warfare: A case study of the French Army in Algeria 1954–1960. *Journal of Strategic Studies, 34*(6), 799–824.

Ouellet, E., & Pahlavi, P. (2015). State actors and non-conventional strategies: A new typology. In S. Kirshbaum (Ed.), *Le système de sécurité occidental face aux nouveaux défis de la sécurité* (pp. 15–48). Brussels: Bruylant.

Pahlavi, P., & Ouellet, E. (2009). Guerre irrégulière et analyse institutionnelle: Le cas de la guerre révolutionnaire de l'Armée française en Algérie. *Guerres Mondiales et Conflits Contemporains, 235*(3), 131–144.

Pahlavi, P., & Ouellet, E. (2012). Institutional analysis and irregular warfare: Israel Defence Forces during the 33-day War of 2006. *Small Wars and Insurgencies, 23*(1), 29–52.

Paparone, C. (2012). *The sociology of military science: Prospects for post-institutional military design.* New York: Bloomsbury Academic.

Perez, C. (2011, March–April). A practical guide to design. *Military Review, 91,* 41–51.

Proctor, P. (2011, March–April). Fighting to understand: A practical example of design at the battalion level. *Military Review,* 91, 69–79.

Schön, D. (1983). *The reflective practitioner: How professionals think in action.* New York: Basic Books.

Scott, R W. (2008). *Institutions and organizations.* Thousand Oaks, CA: Sage.

Smith, E.A. (2002). *Effects based operations: Applying network centric warfare in peace, crisis and war.* Washington, DC: Command and Control Research Program.

Thelen, K. (1999). Historical institutionalism in comparative politics. *Annual Review of Political Science, 2*(1), 369–414.

Thelen, K. (2003). How institutions evolve: Insights from comparative-historical analysis. In J. Mahoney & D. Rueschemeyer (Eds.), *Comparative historical analysis in the social sciences* (pp. 208–240). Cambridge: Cambridge University Press.

United States Army. (2015). ATP 5-0.1 Army design methodology. Washington, DC: Department of the Army.

United States Special Operations Command. (2016). White paper: Design thinking for the SOF Enterprise. Tampa: SOCOM.

Woolgar, S. (1988). *Science: The very idea.* New York: Routledge.

Zweibelson, B. (2015). An awkward tango: Pairing traditional military planning to design and why it currently fails to work. *Journal of Strategic and Military Studies, 16*(1), 11–41.

Zweibelson, B. (2016). Linear and non-linear thinking: Beyond reverse-engineering. *Canadian Military Journal, 16*(2), 27–35.

Part II

New Organizational Forms and Processes

3 Organizational adaptations in the hunt for Abu Musab al-Zarqawi

A review of concepts for analyzing their usefulness[1]

Wilbur J. Scott

General (US Army, retd) Stanley McChrystal's career was unorthodox for one who would rise to the rank of four-star general. His route through the ranks was mostly in elite Ranger units rather than more standard combat brigades. His career also would prove controversial, ending in dismissal for insubordination in 2010 by then-President Barack Obama when McChrystal was commander of Multi-National Forces in Afghanistan.[2] Our interest here is in the first portion of McChrystal's stint as commander in Iraq, from 2003 to 2008, of what was then the US's most elite special operations unit, Task Force 714 (TF-714).

His books, *My Share of the Task* and *Team of Teams*, address this phase of his career. The former is McChrystal's (2014) memoir, Part Two of which has relevance here. McChrystal (2015) wrote the latter with three other collaborators while a senior fellow at Yale University's Jackson Institute for Global Affairs. Subtitled *New Rules of Engagement for a Complex World*, it focuses on the issue of how organizational leaders might adapt to new, befuddling 21st-century milieus. Taken together, the two books provide a remarkable portrait of McChrystal's stumbling but adroit efforts to make sense of an unfamiliar war environment and his corresponding attempts to revise TF-714's structure and culture. This chapter extends his analyses with analytical commentaries on the conceptual usefulness of "lessons learned" for contemporary mission formations.

Distal context

Attacks on the 110-story, twin towers of New York City's World Trade Center and on the Pentagon in Washington, DC, took place on 11 September 2001. Though these compare in significance to the bombing of Pearl Harbor on 7 December 1941 – the official beginning for the US of World War II – there is an important difference. The latter was done by military pilots in planes from the Imperial Japanese Navy. The devastation of 9/11 was carried out by 19 civilian operatives from Osama bin Laden's jihadist think tank, al-Qa'ida. Fifteen were from Saudi Arabia, and the remaining four were from Egypt, the United Arab Emirates, and Lebanon – four countries that are key allies of the US in the Middle East. On the morning of 11 September, they hijacked four civilian passenger planes en route from the US's east coast to California and managed to crash three of them into two of their intended targets. All told, 2,977 Americans died. Eighty-five percent were civilians, 414 were uniformed first responders, and 55 were military personnel.

The US response to the attacks began a month later with a sophisticated air campaign in Afghanistan. This action sought to deprive al-Qa'ida of its training facilities there and to unseat the Islamist Taliban government, dominated by tribal Pashtuns. Within a month, special operations units from the US and Western partner nations, along with non-Pashtun tribal entities, were advancing toward Kabul. In December, the Taliban government collapsed, and Hamid Karzai, a non-Islamist Pashtun, was sworn in as head of a new government. However, the traditional Pashtun area extended across the border into Pakistan and provided the deposed Taliban a sanctuary. A revived Taliban eventually launched an insurgency[3] against the Karzai government.

By then, however, President George W. Bush had ordered the overthrow of Iraq's Saddam Hussein as the next step in what he termed the *Global War on Terror* (cf. Ricks, 2006, Part Two). His war planners, principally Secretary of Defense Donald Rumsfeld, envisioned a short military invasion and a quick departure. It would feature technological wizardry applied with such overwhelming force and sophistication that Iraqi soldiers and assorted camp followers would surrender in *shock and awe*.[4] Deputy Secretary Paul Wolfowitz had no problem with a quick strike but advocated a much longer, guided transition to a fully democratic Iraq. Both men clashed with a State Department wary of starting a war that might upset delicate sectarian balances in the region and of imposing Western democratic values. Secretary Rumsfeld prevailed, so shock and awe it would be.

On 19 March 2003, Tomahawk cruise missiles launched from US planes and carriers shook Baghdad. A lightning-fast invasion followed, and the Iraqi military and Saddam Hussein's government folded as if on cue. Unprepared for the next step,[5] US ground forces stood by as Iraqi celebratory gunfire turned into wanton violence. Looters ransacked museums and government buildings, and high-level staff in most ministries methodically destroyed records, files and equipment to obscure past activities. Asked to comment on whether the disorder was a concern, Secretary Rumsfeld said, famously, "Stuff happens" (McLoughlin, 2003, n.p.). Armed gangs, Sunni and Shi'ite alike, soon assumed responsibility for the security of their own neighborhoods.

A common American perception is that Sunni and Shi'a Muslims have been at each other's throats for centuries, so nasty sectarian strife in Iraq was inevitable after the American takeover. Two-thirds of Iraq's 26 million inhabitants are Arab Shi'ites, 20% Arab Sunnis, and the remaining 12% or so ethnic Kurds, mostly Sunni. Saddam Hussein himself was Sunni and, though not a religious man, he had followed the Arab custom of nepotism by intensely favoring family, tribe and fellow Sunnis. Thus, though Sunnis were a distinct minority numerically, they were fully in charge before the US invasion. Saddam's demise meant Shi'ites would displace Sunnis in positions of power and privilege, perhaps vengefully.

Still, a descent into a vicious civil war was not a given (McChrystal, 2014, pp. 121–2; Allawi, 2007, pp. 1135–8). Iraq's Muslim religious posture in the

20th century was one of toleration, and many Iraqis often thought of themselves as Iraqis rather than simply members of a particular Islamic denomination. For example, a dispute in the 1980s over the Shatt al Arab waterway had led to a bitter 8-year border war with Iran. Iraqi Sunni and Shi'ite conscripts fought in that war side by side against Iran, the dominant Shi'ite country in the region. Hence, strident Sunni views broadcast by ardent Saudi Arabian Wahhabi clerics condemning Shi'ism as *kufr* (heresy) had limited appeal in Iraq. Things did sour after Saddam retaliated against Shi'ites following defeat in Persian Gulf War I in 1992, but there was some basis for optimism.

By May, conditions had deteriorated badly, with essential services from electricity to police protection in near-total disarray. President Bush appointed L. Paul Bremer to preside over Iraq until a new presidential election could be held. Bremer was a diplomatic lightweight, with no experience in the Arab world or with reconstruction issues. Within days of his appointment, he made two momentous but ill-advised decisions: he barred almost all of the 25,000 government workers from Saddam's Sunni Ba'athist Party from any future public sector employment, and he disbanded the entire 400,000-member Iraqi Army, whose officers and elite units were mostly Sunni. The displaced Sunnis quickly became an armed and dangerous threat to the American occupation. Most American units, oblivious of the whys and hows of all this, responded with hard-hitting, often indiscriminate, tactics that angered warring factions and bystanders alike.

Analytical commentary I

Eyal Ben-Ari and his associates[6] (2010) systematically observed Israeli military operations during the second Palestinian uprising – the Al-Aqsa Intifada – from September 2000 to March 2005. What struck Ben-Ari's team is how different the structures and missions of Israeli units in this conflict were from those associated with conventional warfare, and, correspondingly, how inadequate were the concepts military scholars might employ to capture what was taking place. Conventional warfare denotes armed conflicts among state-based, near-peer militaries. World Wars I and II provide well-worn examples. The Al-Aqsa Intifada bore a marked asymmetry: this insurgency pitted informal bands of violence-given Palestinian paramilitaries, rock-throwing youth and civilian suicide bombers against Israeli armed forces, foremost among the world's modern militaries. Palestinian policemen at times joined in the fray against the Israelis. When insurgents violently target civilians, such a conflict may be narrowly denoted, to harken back to President Bush's term, a war on terror, that is, a war with actors who use a certain tactic, *terrorism*.[7]

Wars of this sort typically take conventionally constructed and equipped militaries out of their preferred "big war" modus operandi, forcing them to play a "small war" game more akin to policing.[8] Although these conflicts have military dimensions, they do not have purely military solutions, and how to prevail calls for a different calculus. A hard reliance on military-grade violence often is counterproductive, and occasional bursts of quasi-familiar, warlike violence must be carefully orchestrated. Conventional militaries thus have a mixed track record in nonconventional encounters. Political scientist Seth Jones's (2019) analysis of 181 *insurgencies* showed that those able to attract some combination of combat support, war materiel, money or external sanctuary from outside militaries and patrons were much more likely to prevail than those who had to go it alone. RAND researchers Ben Connable and Martin Libicki (2010) studied 89 high-profile small wars that had occurred since World War II. They found that, when outside militaries intervened directly on behalf of embattled governments, more often than not the big militaries lost altogether, or were forced to make substantial, unwanted concessions, to an inferior military force or to rebels having no formal military at all. Well aware of these pitfalls, Israeli military leaders implemented the unusual mission formations observed by Ben-Ari and his associates.

Following 9/11, the US faced a vastly different security dilemma than did Israel in the Al-Aqsa Intifada and reached different conclusions about how to proceed. Defense analyst Thomas P.M. Barnett (2004, pp. 269–70) summarized the debate among American defense professionals about how to characterize what had taken place on 9/11. Some asserted 9/11 was "only a supercrime", and so the security landscape and mission remained the same: "catch the responsible criminals and life goes on". Others contended: "We only do nation-states. If you need a Taliban or Saddam taken down, fine … [but] our product line should remain the same". Among those who saw a significant shift, there still was caution: "Yes, we must transform Defense as a whole, but we still need to concentrate on defeating other militaries".

Barnett (2004, pp. 121–37) assumed a more macro perspective and drew up a world map depicting two regions. One encompassed countries within globalization's "expanding web of connectivity", the *functioning core*, and the other encompassed areas disconnected from it, the *unincorporated gap*. The functioning core, he argued, posed limited security threats as it is in its countries' interests to favor rule sets encouraging cooperation. Their very *connectedness*, Barnett wrote, promotes "mutually assured interdependence". Notably, the core includes Russia and China, examples of near peers the US's military actually is designed to fight. The unincorporated gap, he continued, is another matter. Its very *disconnectedness* portends danger: countries and actors there lack sufficient incentives to heed violence-avoiding rule sets. In fact, all 140 military actions carried out by the US military on foreign soil between 1990 and 2003 took place in the unincorporated gap.[9] Thus, his broad, long-range recommendation was nonmilitary in scope: *shrink the gap!*

Writing a year later, Barnett (2005, pp. 9–14) assessed the turn of events in Iraq. Shock and awe, he observed, fits well the US military's preferred use of sophisticated, state-of-the-art technologies for waging conventional war. With this advantage, the US *Leviathan*, as he called it, can go it alone in war. But, he continued, it cannot go it alone in small wars or in the seam between war and peace. Here, it needs a *system administrator* that, ideally, would be made up of US troops specifically trained for small wars as part of a multinational force. Barnett's proposal evaporated in the cacophony of conflicting reactions to developments in Iraq.

Proximate context

The American invasion of Iraq attracted a trickle of jihadists and foreign fighters, among them an experienced, hardcore Islamist from Jordan, Abu Musab al-Zarqawi. A former "wannabe" *mujahedeen* (holy warrior) from the war in Afghanistan against the Soviets,[10] he was a long-time bit player in al-Qa'ida. Al-Zarqawi's personal ideology of grisly violence in the name of Islam was extreme even by al-Qa'ida standards, and al-Qa'ida leaders considered him a loose cannon (Warrick, 2015, pp. 7–8). In 2002, he had moved to Iraq on his own in anticipation of the US invasion. His little cell, at first undetected by intelligence sources, went by a name al-Zarqawi had used off and on since 1999, Jama'at al-Tawhid wal-Jihad (Organization of Monotheism and Jihad), or Jama'at for short.

Three sophisticated suicide bombings signaled the disarray in Iraq might be providing cover for a well-thought-out, sinister plot (Filkins, 2003; Filkins & Oppel, 2003; Filkins & Berenson, 2003). On 7 August 2003, a green mini-van parked outside the Jordanian embassy in Baghdad exploded, killing 11 and wounding 65 bystanders. Less than two weeks later, a "gleaming new cement mixer" truck filled with explosives rammed the Canal Hotel, where the United Nations' Baghdad headquarters was located. Among the 22 dead were the head of the UN mission in Iraq, Brazilian Sérgio Vieria de Mello, and top members of his staff. And, on the first day of Ramadan that October, five suicide bombers, within a 45-minute period, drove cars into the headquarters of the International Committee of the Red Cross and four Iraqi police stations, killing 34.

These attacks, taken together, did not escape the notice of US intelligence. Someone was targeting organizations who had the niche expertise to assist in Iraq's reconstruction. In January of 2004, an analyst with TF-714 approached his commander. "Sir", he said, "we have good reason to believe Zarqawi is in Iraq. And, sir … he's building up a network" (McChrystal, 2014, pp. 112–3).

TF-714 itself was an outgrowth of a mission gone awry in 1980, *Operation Eagle Claw*. It had attempted the rescue of 53 American hostages held in Tehran. An ambitious, multi-stage mission, it was carried out by elite units drawn from all four branches of the US military. Unaccustomed to working with each other, the mission ended catastrophically when a Navy helicopter and Air Force C-130 aircraft collided in the Iranian desert, killing five airmen and three marines. The failure led to the creation of the multi-service, joint Special Operations Command (SOCOM). The TF-714 McChrystal joined 23 years later combined its headquarters in Fort Bragg, North Carolina, with Army Delta Force ("Green"), Navy SEAL ("Blue") and Army Ranger elements in Iraq and Afghanistan. By almost any yardstick of efficiency, TF-714 was among the most finely honed military units in the world. TF-714's initial mission in Iraq had been the capture or killing of "the deck of cards", a set of playing cards created by the US Department of Defense and embossed with pictures of the most-wanted members of Saddam Hussein's network. By late 2003, TF-714 had already worked through most of the deck.

Meanwhile, al-Zarqawi was poised for the next phase of his plan: fomenting all-out civil war between Iraq's Sunnis and Shi'as. In January of 2004, a Kurdish

Peshmerga[11] patrol arrested a Pakistani al-Qa'ida operative carrying a thumb drive containing a letter from al-Zarqawi to Osama bin Laden. Al-Qa'ida had issued a formal admonishment of Jama'at for its suicide bombings in which Muslim bystanders were killed. Al-Zarqawi wrote the letter to explain his strategic thinking and ask al-Qa'ida for its public endorsement. In it, he dismissed the Americans as "an easy quarry, praise be to God". His concern and hatred were directed toward Shi'ites:

> [They are] the insurmountable obstacle ... the crafty and malicious scorpion ... a sect of treachery and betrayal throughout history. ... I come back and again say that the only solution is for us to strike the religious, military, and other cadre among the Shi'a with blow after blow until they bend to the Sunni.
>
> (Allawi, 2007 p. 233)

Six weeks later, multiple suicide bombers from Jama'at blew themselves up among thousands of Shi'a worshippers in the city of Kerbala. At least 169 died, and hundreds were wounded. Shi'ite militias surged into Sunni neighborhoods for revenge and, as Sunnis retaliated, full-scale violence between the two groups erupted over the following months. The American command assigned TF-714 a new primary mission: capture or kill al-Zarqawi.

TF-714 knew the opaque outlines of al-Zarqawi's game plan, and what they did know suggested a worrisome and increasingly complex insurgency. On one level were Sunni dissidents concentrated in the so-called Sunni Triangle, an area bounded by the cities of Ramadi and Fallujah in Anbar Province to the west of Baghdad, Baghdad itself and Saddam Hussein's hometown of Tikrit to the north. Their grievances sprung from their total displacement by the newly installed Shi'ite-run government, military and police forces, a dilemma for which many Sunnis blamed the Americans as well as the Shi'ites. Jama'at activities directed by al-Zarqawi in Anbar Province added an edge of religious fervor to the Sunni anger. On the second level, Jama'at suicide bombings had aggravated tensions between Sunnis and Shi'ites to a boiling point. To render a breakdown in all spheres of public life, al-Zarqawi wanted this phase of Iraq's plight to be in the hands of the violent extremists. US intelligence indicated that the suicide bombings were not being carried out by local Sunnis but by volunteer foreign fighters guided by Jama'at.

McChrystal did an assessment. The average age of his operators was 35; they were all sergeants or officers, no privates here. They were strong-willed, mature, highly professional and very opinionated. "Given a worthy mission – an idea they could come to believe in", McChrystal observed, "they committed to it unlike anything I had yet seen in my military career." Their traditional specialty was doing surgical, direct-action missions – usually dangerous and complicated hostage rescues or pinpoint assaults – under the label of the 3Fs: *find, fix, finish*. However, there was room for improvement. They tended to be a "tribe of tribes". He wrote:

> Stratification and silos were hardwired throughout the taskforce. Although our units resided in the same compound, most lived with their "kind" ... used

different gyms ... controlled access to their planning areas, and each tribe had its own brand of standoffish superiority complex. Resources were shared reluctantly. Our forces lived a proximate ... parallel existence.

(McChrystal, 2015, pp. 120–3)

These silos made it easy for something to fall between the cracks. During an early inspection of TF-714 facilities at Baghdad airport, McChrystal (2014, pp. 105–8) opened a closet door in one hut and found it piled high with plastic and burlap bags filled with material picked up on TF-714 raids – wallets, cell phones, personal items, military gear and the like. He was told there had not been time to go through it all. Here, McChrystal observed, was a "blink", a bottleneck. TF operators in the field picked up potentially intelligence-rich material, but it had to be screened by professionals in a rear area before being sent back to the US for centralized analysis. It could take weeks for the cycle to complete if it did at all.

To address this problem, McChrystal felt, all agencies and units who had a hand in processing relevant intelligence need be included in the loop. This would mean bringing together representatives from the National Security Agency (NSA), the Central Intelligence Agency (CIA), the Federal Bureau of Investigation (FBI) and the intel sections of other military commands. McChrystal clearly did not have the authority to arrange this himself. To do so, he had to go up levels of command to General John Abizaid, then commander of the 27-country central region for all US military operations. He convinced General Abizaid to set up a grouping of representatives from relevant intelligence units to process the raw evidence for TF-714 in one location, at the American airbase in Balad, Iraq.

This bringing together of intelligence representatives in one location allowed TF-714 to more efficiently identify and search out key Jama'at personnel. In the summer of 2004, TF-714 carried out up to 18 raids a month. However, al-Zarqawi himself avoided detection and continued to direct attacks. Puzzled, McChrystal (2015, pp. 13–7) revisited what his intel people knew about the organizational structure of Jama'at. Rather than being hierarchical in design, it appeared to consist of horizontal nodes. The adroit use of old and new communication technologies allowed them to "scoot and shoot" without patterns allowing easy detection and identification. The style seemed to follow a dictum of *centralization of decision but decentralization of execution*, that is, al-Zarqawi and two or three close confidants decided what to do and when, and cells in varying locations carried out their assignments even as unforeseen developments popped up. The posting of homemade videos on the internet promoted their brand – infamously, of grisly beheadings or suicide bombings – and attracted an influx of 100–150 foreign fighters per month. Somehow, Jama'at integrated them rather seamlessly into cells across Iraq.

McChrystal was frustrated. His organization was already highly efficient before he arrived, and he had doubled its efforts to become even more efficient. On the other hand, he would not describe Jama'at as efficient. It was woefully resourced and only loosely organized. But it was, McChrystal had to admit, highly adaptable and, even if somewhat hit-and-miss, managed to carry out its plans with disconcerting regularity. Then it hit him: the *new war's environment* favored *flexibility over efficiency*. He did not need TF-714 to become more efficient; he needed for it to become more flexible, more adaptable (McChrystal, 2015, pp. 80–2).

Meanwhile, al-Qa'ida's top leadership also had become sufficiently impressed. In October of 2004, bin Laden gave al-Zarqawi his blessing, but with an admonition: kill more Americans, fewer Muslims. Jama'at was now officially the al-Qa'ida Organization in the Land of Two Rivers, or simply, *al-Qa'ida in Iraq* (AQI).

Analytical commentary II

Small wars have been around for centuries – note America's own Revolutionary War of the 1770s in which its colonies fought for independence from Great Britain. The colonies' crude militias and the nascent military fought it out with the renowned British Navy and established British Army. Outfitted in fancy black hats, red jackets and white trousers, the British Army – the redcoats – liked to line up in open fields with muskets and exchange gunfire point-blank with another army doing the same. Lore has it the overmatched colonists wore earth tones for camouflage and used trees for cover and concealment while taking potshots at the chagrined redcoats.

However, the 21st century has produced a different variant, if not category. The person most responsible for applying the label "new" to 21st-century wars is political scientist Mary Kaldor (1999). In her thinking, a particular contextual variable, *globalization*, gives new wars their distinctive features. Rapidly accelerating technological innovation and cross-national trade and investment have created a new, wholly independent system of global capitalism with its own brand of winners and losers. Among the losers, she argues, are authoritarian states whose internal controls over heterogeneous subpopulations rely upon isolation and who, therefore, have become destabilized by this opening up of the world. Most important in her view is how all this has affected certain states' monopoly over organized violence. Kaldor's new wars are fought by a variety of non-state actors, and perhaps elements of the decaying state itself, all bankrolled by predatory private financing – loot and pillage, black-marketeering, strong-armed informal taxation, diaspora support and the like. Motivations often are rooted in identity politics – notions of ethnic or religious communities – and its groups seek control through polarization and exclusion, ethnic cleansing and the geographic displacement of out-groups.

Despite not embracing the meaning of new wars in exactly the way Kaldor does, Ben-Ari and his associates (2010, pp. 30–1) are persuaded there *is* something fundamentally new about all wars in the 21st century. Contemporary small wars have a critical, exaggerated public relations dimension not typically associated with past wars. They note, for instance, the growing transparency of contemporary warfare through *global surveillance* – a term suggested by military sociologist Martin Shaw (2005, p. 75) – made possible by the internet and social media. Actions that in the past would have been privy to those who were physically present might now be viewed in real or delayed time by interested parties located throughout the world. One result is that every move, every encounter, every casualty has the potential to be magnified for political purposes. This can prove hugely advantageous or detrimental to military leaders and other actors, depending upon how astutely aware they are of the potential.

In this vein, McChrystal (2014, pp. 122–4), early in his command, convened a conference at Bagram airbase in Afghanistan for his TF-714 commanders and senior enlisted personnel. Required reading for day one was *Modern Warfare* by French military officer Roger Trinquier (1985), based on his experiences in Indochina and Algeria. The small, 88-page book was originally published in 1961, so it long predated the internet and its global capacities. Of relevance still were Trinquier's incisive descriptions of small wars with terrorist insurgents and his ideas on how to confront them successfully, including the tailoring of military actions to address the political dimensions of the conflict. McChrystal intended discussions to raise awareness among his leaders and bring their opinions out in the open, but also to make his own positions clear. Day two of the conference was devoted to the viewing and discussion of the 1966 movie *The Battle of Algiers* (*La Bataille d'Alger*). The film depicts the French 10th Parachute Division's systematic use of torture in taking over of the casbah quarter of Algiers city in 1957, a policy Trinquier endorsed. McChrystal explained his disagreement and made clear that TF-714 would not be emulating this portion of Trinquier's recommendations in its handling of detainees.

Ben-Ari and his associates (2010, pp. 53–62) explored other ways Israeli military leaders have assessed and adapted to *global surveillance*. For starters, the constant mingling of soldiers and civilians already had inspired discourse on human rights. From this, the Israeli Defense Forces (IDF) had adopted a code of ethics focusing on "the dignity of man", that is, principles to be followed in decision-making and behaving when encountering civilians, using violence and treating prisoners. A more recent adaptation was the creation of *in-between organizations* to manage troops' relations with Palestinians and other outside groups. During the Al-Aqsa Intifada, for example, *humanitarian officers* were attached to IDF units to serve as arbitrators in sensitive interactions between Israeli soldiers and Palestinian civilians. Similarly, the IDF allowed the organization Machsom Watch to provide women volunteers as observers and intermediaries at troublesome checkpoints. In-between measures were designed to elevate troops' behavior, reduce the likelihood of events going awry and increase transparency for the world outside the IDF.

A global-surveillance capacity does not guarantee increased humanitarian concern. Al-Zarqawi's AQI, for instance, often had operatives follow suicide bombers to video-record the gory aftermaths. Along with videos in which al-Zarqawi himself beheaded individuals held prisoner by AQI, such recordings were distributed on the internet to laud strategic accomplishments and recruit fighters from outside Iraq to join the cause. Before the advent of technologies making possible the instantaneous worldwide distribution of such materials, achieving the desired notoriety would have been much more difficult. Al-Zarqawi's "genius" was to envision how to exploit today's interconnected world for his own nefarious purposes.

Building a team of teams

Another theme in the Bagram conference was the dire need for deep intelligence. *Modern Warfare* and *The Battle of Algiers* had laid bare the outright ignorance of the French about their adversaries in Indochina and Algeria. To underscore TF-714's shortcomings here, McChrystal (2014, p. 123) had pointed toward the compound's perimeter and declared, "We fundamentally do not understand what is going on outside the wire". The Green commander suggested one remedy: a revision of the 3Fs to find, fix, finish, *exploit, analyze*. The latter two steps emphasized how each mission should consciously seek intelligence to lay bare specific details of AQI's network.

McChrystal already had brought representatives from relevant intelligence agencies together with fellow TF-714 team members at Iraq's Balad airbase. However, these intel analysts, like his own special operators, had come from agencies whose cultures were exclusive rather than inclusive. Their tendency was to hold information close to the vest and to share it only sparingly with outsiders. In part, this reflected the nature of doing "secret" work: *secret* or *top-secret* clearance provided a person access to highly sensitive intelligence only if he or she also had an official *need to know*. The default position among intel personnel was to assume that others did *not* need to know. These tendencies were reinforced by the living and working conditions at Balad airbase. McChrystal wrote:

> The blinks were even worse between the Task Force and our partner organizations from the CIA, FBI, NSA, and conventional military units. Initially, representatives from these organizations lived in separate trailers, with limited access to our compound. [I]n the name of security, these physical walls produced mis-information and distrust. ... We needed everyone to know *someone* on every other team, so when they ... had to work with the unit that bunked next door or their intel counterparts in D.C., they envisioned a friendly face rather than a competitive rival.
>
> (McChrystal, 2015, pp. 122–3)

McChrystal (2015, pp. 128–30) and his staff wanted the emotional ties and adaptive capacity found within TF-714's individual units to extend throughout the task force. They were not thinking everyone should become proficient at all tasks. Green and Blue teams are much better at what they do than intel analysts ever would be, and vice versa. What they had in mind, they decided, was "to fuse *specialized expertise* with *generalized awareness*". Everyone would need a holistic understanding of all the moving parts. The more nuanced the understanding of the big picture, the more a person anywhere in the task force would be able to provide meaningful input on the fly. They labeled this end state *shared consciousness*.

To achieve this, they concluded, some relationships between individuals on the constituent teams needed to resemble those within a team. This would require getting personnel out of their silos and into a common interactive space.

McChrystal turned his attention to their Soviet-era central building at Balad airbase. He had base engineers gut its warren of small offices and cubicles and replace them with an open floor plan. The resultant expansive area became the Joint Operations Center, where a bank of video screens displayed live updates of all operations in Iraq, including feeds from overhead military drones.[12] A U-shaped arrangement of portable tables and chairs in the middle was for McChrystal, his staff and unit heads, who now could all see each other and talk directly.

Of course, an alteration in a physical arrangement in itself does not guarantee a change in culture. A central feature of TF-714, the daily operations and intelligence (O&I) briefing, provided a unique opportunity. The briefing was usually a video-teleconference involving a handful of leaders or their representatives from headquarters in Fort Bragg, a few offices in Washington, DC, and the biggest bases in Iraq and Afghanistan. To encourage a culture of sharing, McChrystal envisioned a much more extensive Skype session. He took the unusual step of announcing that everyone connected with TF-714 and its extended network had *the need to know*, and, hence, all TF members and affiliates, located at Balad or elsewhere in the world, were welcome to "sit in". Anyone, McChrystal (2015, pp. 138–9) wrote, "regardless of their position in the org charts' silos or tiers could glance up at the screens and know instantly about major factors affecting [their] mission at that very moment".

Although everyone was invited to sit in, not everyone was required to do so. However, the daily brief caught on as *interactive discussion* became the norm. O&I briefings now generated excitement and became a popular "hangout" for personnel throughout the network (McChrystal, 2015, pp. 167–9). The time difference meant that morning in Fort Bragg and Washington, DC, was 1600 hours (4 p.m.) in Iraq and Afghanistan. Green, Blue and Ranger operators getting ready for that night's raids, who in the past had not themselves been party to this briefing, could now personally get the most up-to-the-minute information; intelligence personnel, who in the past would have passed their input to an anonymous network, now could see their briefs having immediate tactical relevance. Further, they could talk face-to-face, even if some were doing so via live video-feeds. Participants found they could query each other about details and procedures, analysts from different silos who had reached divergent interpretations could debate the meaning of what was found, and operators in different areas could share observations about eerily similar enemy tactics in geographically separated areas.

McChrystal (2015, pp. 214, 217–8) also revised his own leadership style from "hands-on" to "eyes-on, hands-off". Regulations required that he personally approve strikes on targets when American troops were not in live-fire contact with the enemy. As his confidence in the *shared consciousness* of his teams increased, he pushed the authority to make such decisions far down the chain of command. His basic guideline was, "If something supports our effort, as long as it is not immoral or illegal, you can do it". His initial rationale was that a "70% solution right away" was better than a "90% solution later". He soon observed the reverse to be true: in the climate of *shared consciousness*, a decision on the fly by engaged subordinates typically produced a 90 percent solution.

These changes produced a flow of detailed information. For instance, TF-714 confirmed that AQI had only a very small number of hardcore members, maybe 100 or 200. However, they were able to vastly multiply their operational reach by tapping into extensive, pre-existing networks of Sunni tribal affiliations, extended families, business enterprises – some legal, some not – militias, assorted other groupings and even bystanders. Some of these cooperated, or at least enabled AQI by tolerating their presence, because of discontent with the Shi'a power grab or the American presence. Some had grievances with Shi'ites or Americans, having lost family members in encounters with them. Some did it for excitement and adventure. But, significantly, many did so out of fear because of AQI's ruthless use of intimidation and physical violence against Sunnis who might stand up to them. In any case, the key for TF-714 lay in developing initiatives to split up, pre-emp, co-opt or otherwise exploit each of these on its own terms. And it implied a big "not" – do not treat them all as if they were AQI.

They also began to see how AQI so seamlessly integrated foreign fighters into its operations. There was a discernible pattern (McChrystal, 2014, pp. 170–3).[13] Most foreign fighters were young Muslim men, heretofore politically inert and only mildly religious, who had become radicalized in their home countries after viewing Islamist materials and videos on the internet. Portrayals of the abuse of Muslim prisoners at Abu Ghraib often were a "first jolt to action". These now-awakened youth typically gravitated toward a local mosque, where AQI "spotters" befriended and directed them to local cells. There, they became immersed in AQI's version of jihad and the war in Iraq. For most, it went no further than this, but a small number proceeded to pledge total allegiance to the cause. These were flown to Damascus, Syria, where AQI handlers divided them into task groups. The groups (and munitions) were funneled into what TF-714 called "ratlines" – a system of "safe-houses" through which they were escorted to the Syrian border and then into designated locations in Iraq. Usually, the first time a young suicide bomber saw Iraqis outside the AQI network was just before he killed them.

Analytical commentary III

Research by military scholars during and since World War II has driven home the point that *unit cohesion* binds individual soldiers into solid fighting groups. Their studies, largely of conventional military units, have shown how continuous face-to-face interaction with fellow soldiers under arduous circumstances creates emotional attachments which, when aligned with organizational goals, produce military units able to persevere in the worst of combat conditions. Ben-Ari and his associates (2010) have raised the question of whether this contention holds for very irregular mission formations in very unconventional settings. Their answers are a qualified "yes" – some variants of *cohesion* and *organizational integration* are necessary – and "no" – these are not produced in the same way as in *textbook units* operating in conventional settings. The research cited above would predict a loss of military effectiveness.

A hallmark of irregular mission formations is their fragmentation and recombination to meet unusual conditions, resulting in potentially tenuous *loose couplings* among constituent parts. McChrystal verbalized this situation by likening TF-714 in his early days to a "tribe of tribe." Although there was substantial *internal cohesion* within its subunits, loose couplings inhibited interaction across subunits and interrupted feedback loops, to the detriment of the overall mission. Ben-Ari and his associates noted a potentially similar situation during a visit to the dining hall of a specially constructed Israeli unit during the Al-Aqsa Intifada:

> On the right [row of tables] are two tank crews with red eyes and unshaven, after having spent the whole night in observation. Opposite them and much more rowdy are about fifteen infantry soldiers from an infantry company. Crews of anti-aircraft units sit at a different table, as do female soldiers who are in charge of some surveillance instruments. … It seems everyone is aware of who is in the dining hall but the social communication takes place primarily within the groups.
>
> (Ben-Ari et al., 2010, p. 73)

Still, such *instant units* may endure. TF-714, after all, was highly efficient – it just was not flexible enough. One possible solution, attempted by McChrystal and his staff, might be to raise the shared consciousness among subunits by increasing the interaction capacities at the very nodes where loose couplings do connect them. They accomplished this by restructuring some of the task force's activities – in this case, O&I briefings – so that representatives from each subunit now had occasion to interact face-to-face in the line of duty on a regular basis. These opportunities were extended throughout TF-714 and its affiliates to geographically dispersed representatives through real-time video-teleconferencing. Apparently, a video presence, or maybe even immersion in virtual realities, has sufficient "carrying capacity" to produce intended effects. One can envision even greater diffusion of participation by using cell phones with encrypted capabilities.

Ben-Ari and his associates (2010, pp. 76–81) had speculated that such or similar moves might enhance what McChrystal called shared consciousness through two mechanisms. One, they suggest military members over time may develop the capacity to cohere easily, not just to one particular group or unit, but to most any military grouping as a more general ability. This would promote the capacity for *swift trust*, thereby allowing heightened levels of cohesion to occur very quickly and through limited interactions. Thus, they suggest (2010, p. 86), soldiers faced with a new situation might "proceed under the assumption that the proper way to act is to cooperate, take others into consideration, and be identified and committed to the new social framework".

Two, following an earlier analysis by military sociologists David Segal and Meyer Kestenbaum (2002), Ben-Ari and his associates (2010, pp. 77–8) underscore their distinction between *unit cohesion* and *esprit de corps*. As indicated above, cohesion usually is portrayed as an emotional attachment to others resulting from enduring face-to-face interaction; esprit de corps denotes an emotional attachment arising from identifying oneself as a member of a larger "imagined community" (cf. Anderson, 2006). Of course, both cohesion and esprit de corps may develop within the same group context. Indeed, this could align the aims of the small group with those of the larger organization in which it is embedded; where this not so, cognitive dissonance might be an issue for members of the small group. Segal and Kestenbaum point out another interesting possibility: in the absence of sufficient cohesion, esprit de corps might serve as an alternative source of positive motivation and emotional support.

Concluding observations

On 7 June 2006, two 500-pound, laser-guided bombs from a US Air Force F-16 jet killed al-Zarqawi in a remote farmhouse as he met there with his spiritual advisor. A highly interconnected team – two and a half years in the making – of special operators, both on the ground and in the air, intelligence analysts with expertise in every conceivable form of information, and skilled support personnel had combined to make that day possible. McChrystal summarized it this way:

> The organization that pulled this off was worlds apart from the one [I took over] in September 2004. Driven by necessity to keep pace with an agile enemy and a complex environment, we had … fused a radical sharing of information with extreme decentralization of decision-making authority. In doing so, we had a structure unlike any force the US military had ever fielded. Gone were the straight lines and right angles of a traditional … org chart; we were now amorphous and organic, supported by crisscrossing bonds of trust and communication that decades of managers had labeled as inefficient, redundant, or chaotic.
>
> (McChrystal, 2015, p. 242)

In keeping with his adapted leadership stance, neither the decision to track the particular suspects that led to al-Zarqawi's exact location that day nor the decision to launch the F-16 strike were made by McChrystal. Those were made by an intel sergeant and one of his unit commanders, respectively. Military psychologists Uzi Ben-Shalom and Eitan Shamir (2011) have explored this leadership style, designated *mission-command* in the context of the IDF's new mission formations. Because it allows for spontaneous decision-making by individuals down the chain of command, effective mission-command requires both a thorough understanding of, and commitment to, mission objectives and the top commander's *intent*. As such, it may indeed facilitate a high degree of flexibility, but also incurs a certain amount of risk. McChrystal took that risk.[14]

There is, too, a larger lesson to be learned about prevailing, or not, in small wars against terrorism. Israeli military analyst Kobi Michael (2018) has warned of the deceptive lure reliance on special forces operations may hold for policy-makers. Specifically, successful special forces missions, surgically performed and widely celebrated after the fact, may impart a false sense of substantial progress – a "silver bullet" – in otherwise messy small wars. In fact, the killing of al-Zarqawi did not end AQI's threat or prevent the transformation of that small group later into something much larger and more malicious.

For the first 18 months though, things did appear promising, with a fortuitous turn of events in Anbar Province. Initially welcomed by many Sunnis as an anti-dote to Shi'a violence and the American presence, AQI by this time had badly overplayed its hand there. Its brutal intimidation of Sunnis who questioned it, its stark version of Sharia law – weird to less-fanatically religious Iraqis – and its disinterested mismanagement of city services had led some Sunnis to consider the American presence a lesser evil (Totten, 2007).

The 1st Armored Division's 1st Brigade arrived in Ramadi at this oppor-tune moment. Brigade commander Colonel Sean MacFarland, reflecting a very unusual mindset for the senior officer of an armored unit,[15] set as his two top priorities the protection of local Sunni kingpins (sheiks) from reprisal and the promotion of round-the-clock security by an indigenous Sunni police force. One of MacFarland's staff officers, a former special forces operative and Arabic-speaker, Captain Travis Patriquin, soon developed a close relationship with one disaffected sheik, Abdul Sattar Abu Risha, the first to defect from AQI's orbit of control. Ultimately, fifty-some sheiks joined in a movement the sheiks them-selves called Sahawa al-Anbar, the Anbar Awakening (McChrystal, 2014, pp. 240–2). MacFarland convinced the sheiks to reassert control by sending their own private militias for formal training at the Iraqi police academy. In 2007, their informal militias thus became the official uniformed police forces of Anbar Province.[16]

Meanwhile President Bush, signaling a new policy direction in Iraq, had sacked Secretary Rumsfeld and appointed Robert Gates as his new Secretary of Defense. Under Secretary Gates, General David Petraeus became the new com-mander of Middle East operations in 2007. He installed a new strategic doctrine for the wars in Iraq and, later, in Afghanistan. Petraeus also received authorization for an initiative referred to as "the surge" (cf. Ricks, 2009, Part Two). It added 20,000 American ground troops to the 150,000 or so already in Iraq. Petraeus characterized its strategic importance, however, as a shift in focus from capturing and killing insurgents, that is, *counterterrorism*, to securing the physical safety and eliciting the support of the Iraqi population in resisting violent extremists, that is, *counterinsurgency* (COIN).

McChrystal (2014, pp. 242–6) already had set up a special subunit in TF-714 to make the most of the Awakening. A British Special Air Services officer on loan to him, Lt Colonel Graeme Lamb, headed this subunit. Lamb, well steeped in COIN theory and practice, called his mission *strategic reconciliation*. His team put their listing of Iraq's violent groups on a continuum, with the most violent at the two ends and those less so toward the middle. At one end was Sunni-based AQI, and at the other were Shi'a-based Iranian proxies. Those at the very ends of the continuum were assumed, for the time being, to be *irreconcilables*, with every other grouping designated *potentially reconcilable*. Lamb proposed beginning with the groups just shy of the extremes. Their leaders would be accosted and given realistic options for assisting the "new Iraq". They did not have to join forces with the Shi'ite-dominated government in Baghdad or the Americans – they needed only agree not to align with the irreconcilables. Lamb himself led the negotiations. "We can offer them a way out, we can show them daylight, yeah", he said, "but if they don't take it, we can put 'em in the fucking grave" (McChrystal, 2014, p. 246).

The surge centered on Baghdad and its surrounding areas. At first, death counts rose among Iraqi civilians, policemen and US soldiers. However, by the end of the year, the surge's military and peace-promoting tactics, the Awakening's influence in the Sunni Triangle and a cease-fire order to Shi'ite militias by the influential Shi'a cleric Muqtada al-Sadir all served to dramatically reduce violence across Iraq. The US began troop reductions in 2008, and President Bush, at the request of Iraqi Prime Minister Nouri al-Maliki, signed a status-of-forces agreement guaranteeing withdrawal of all US combat troops from Iraq by 31 December 2011, a timetable upheld by President Barack Obama.

Despite these gains, Prime Minister al-Maliki, leader of the Shi'ite-based Islamic Dawah Party, staunchly resisted pressure from the US to share power with Sunnis, especially in the Triangle, and Kurds in northern Iraq. Emboldened, remnants of AQI targeted several top Sunni sheiks for assassination. When Maliki's government went so far as to suspend permanently the paychecks for the Sunni police forces in Anbar, Sunnis en masse abandoned the Awakening, and large numbers of frustrated Sunnis once again saw AQI as an option for striking back. A revitalized and much enlarged AQI morphed through several stages in western Iraq and eastern Syria to emerge in 2011 under a new title: the Islamic State of Iraq and Syria (ISIS).

Notes

1   The views expressed in this chapter are solely those of the author and do not necessarily reflect positions, official or otherwise, of the US Air Force Academy, US Air Force or the US Department of Defense.

2   In April of that year, McChrystal and his staff had entertained a reporter from *Rolling Stone* magazine for several days at a Paris conference and trip to Berlin, which included bouts of drinking together. The resulting story cited derogatory remarks by them about the commander-in-chief, vice-president, the president's national security advisor and the ambassador to Afghanistan (Hastings, 2010). McChrystal right away admitted he had "messed up". At his retirement ceremony, he stated remorsefully, "my career did not end as I would have wished", but Secretary of Defense Robert Gates assured those gathered that McChrystal left "with his place secure as one of America's greatest warriors" (Braumiller, 2010).

3   An *insurgency* most generally is a violent challenge mounted by non-state actors against an existing government. A slightly more comprehensive definition is "a political and military campaign waged by a nonstate group (or groups) to overthrow a regime or secede from a country" (Jones, 2017, p. 7). See also O'Neill (2005) for a taxonomy of insurgencies.

4 See Harlan Ullman and James Wade (1996) for a treatise on *shock and awe* as strategic thinking and as a tactic.

5 Although dissension among Defense and State Department officials stymied preparation for post-invasion contingencies, the US did have a transition plan of sorts. Retired Lt General Jay Garner had been selected to draw up transition and reconstruction provisions 2 months before hostilities commenced. Garner previously had directed humanitarian relief efforts for the Kurdish minority after Persian Gulf War I. His hastily assembled taskforce busied itself with how Iraqi resources might be tapped to restore any disrupted services and to handle large numbers of displaced persons. However, the exact authority and support he would have once on the ground remained vague.

6 At the time of this research, the investigative team included Eyal Ben-Ari, Professor of Sociology and Anthropology, Hebrew University of Jerusalem; Zeev Lehrer, formerly head of research for the Chief of Staff Advisor on Gender Issues of the IDF; Uzi Ben-Shalom, head of research for the Military Psychology Center at the Ground Forces Command, IDF; and Ariel Vainer, Military Psychologist for the Ground Forces Command, IDF.

7 *Terrorism* may be defined as the unlawful use of violence and intimidation against civilians and other noncombatants for political gain. Under Article 3 of the Geneva Convention, *combatants* are members of armed forces, except for medical and religious personnel. Combatants may legally use violence in armed conflicts against enemy combatants and also are legal targets themselves.

8 Because of the asymmetry in military might, insurgencies usually use a guerilla strategy in confronting a formal military. Insurgencies may turn to conventional warfare when they enjoy a military advantage, usually because they have been more fully equipped by an external military or because the military they are fighting has become drastically weakened (Jones, 2019, Chapter 3).

9 These actions included evacuations, peace or relief efforts, shows of force, contingency positionings and actual combat actions, with four areas – Haiti, Somalia, the former Yugoslavia and Iraq – making up 80% of the Pentagon's business. In the first three, faltering states became incubators for terrorism as local gangs and militias, overlapped with kinship, ethnic or sectarian groupings, stepped in to perform tasks ordinarily provided by effective governments and economies. Such developments were especially likely, Barrett noted, in areas experiencing heavy doses of Westernization but limited economic development.

10 Al-Zarqawi volunteered to join the fight against the Soviets in Afghanistan, but actual combat had ended there by the time he reached the front.

11 The Kurdish Peshmerga are the official military forces of the Kurdish region of Iraq. They, not the Iraqi military, were responsible for the security of that region of Iraq.

12 The Army refers to military drones as *unmanned aerial vehicles*. The Air Force, which actually "flies" the drones used operationally in Iraq and Afghanistan – MQ-1 Predators and MQ-9 Reapers – designates them *remotely piloted aircraft* (RPAs). RPA pilots are located at several sites in the US, most famously at Creech Air Force Base to the northwest of Las Vegas, Nevada. The RPAs themselves are housed, maintained and launched from sites in the Middle East. The connection between pilots and RPAs is made using satellite links.

13 For more detailed descriptions and analyses, see the work of Marc Sageman (2004, 2008). A former CIA field officer, Sageman also holds MD and PhD degrees in forensic psychiatry and political sociology. His books, and accounts of terrorist recruitments and networks, are based on data sets he himself constructed of al-Qa'ida operatives.

14 See Anthony King (2017) for a nuanced analysis of *mission-command* and its application in the leadership styles of McChrystal and of Marine General James Mattis.

15 Colonel MacFarland had studied an earlier, successful campaign by the 3rd Armored Cavalry Regiment led by Colonel H.R. McMaster to retake the city of Tal' Afar in 2005.

That operation had been hailed by some as a model for dealing with Iraq's entrenched insurgencies (cf. Packer, 2006; McCone et al., 2008; Scott et al., 2009).

16 Tragically, Captain Patriquin (along with Marine Corps Major Megan McClung and Army Specialist Vincent Pomante) was killed by an improvised explosive device (IED) in December of 2006. Sheik abu Risha was killed in June of 2007 by an IED that appeared to have been set specifically for him.

References

Allawi, A.A. (2007). *The occupation of Iraq: Winning the war, losing the peace*. Yale University Press.

Anderson, B. (2006). *Imagined communities: Reflections on the origin and spread of nationalism* (revised ed.). Verso.

Barnett, T.P.M. (2004). *The Pentagon's new pap: War and peace in the twenty-first century*. G.P. Putnam's Sons.

Barnett, T.P.M. (2005). *Blueprint for action: A future worth crafting*. G.P Putnam's Sons.

Ben-Ari, E., Lerer, Z., Ben-Shalom, U., & Vainer, A. (2010). *Rethinking contemporary warfare: A sociological view of the Al-Aqsa intifada*. State University of New York Press.

Ben-Shalom, U., & Shamir, E. (2011). Mission command between theory and practice: The case of the IDF. *Defense and Security Analysis, 27*(June), 101–117.

Braumiller, E. (2010, July 23). McChrystal ends his service with regret and a laugh. *The New York Times,*. Retrieved from www.nytimes.com/2010/07/24/us/24mcchrystal.html

Connable, B., & Libicki, M.C. (2010). *How insurgencies end*. RAND, National Defense Research Institute.

Filkins, D. (2003, August 7). At least 11 die in car bombing at Jordan's embassy in Baghdad. *The New York Times*. Retrieved from www.nytimes.com/2003/08/07/internat ional/worldspecial/at-least-11-die-in-car-bombing-at-jordans-embassy.html

Filkins, D., & Berenson, A. (2003, October 28). The struggle for Iraq: Suicide bombers kill at least 34. *The New York Times*. Retrieved from www.nytimes.com/2003/10/28/wor ld/the-struggle-for-iraq-insurgency-suicide-bombers-in-baghdad-kill-at-least-34.html

Filkins, D., & Oppel, R.A. Jr. (2003, August 20). Huge suicide blast demolishes U.N. headquarters in Baghdad. *The New York Times*. Retrieved from www.nytimes.com /2003/08/20/world/after-war-truck-bombing-huge-suicide-blast-demolishes-un-headq uarters-baghdad.html

Hastings, M. (2010). The runaway general. *Rolling Stone*. Retrieved from www.rollingston e.com/politics/news/the-runaway-general-20100622

Jones, S.G. (2019). *Waging insurgent warfare: Lessons from the Vietcong to the Islamic State*. Oxford: Oxford University Press.

Kaldor, M. (1999). *New and old wars: Organized violence in a global era*. Palo Alto, CA: Stanford University Press.

King, A.C. (2017). Mission Command 2.0: From the individualist to the collectivist model. *Parameters: The U.S. Army War College Quarterly, 47*(Spring), 7–20.

McChrystal, S., General, US Army, retd (2014). *My share of the task: A memoir*. Portfolio-Penguin.

McChrystal, S., with Collins, T., Silverman, D., & Fussell, C. (2015). *Team of teams: New rules of engagement for a complex world*. Portfolio-Penguin.

McCone, D.R., Scott, W.J., & Mastroianni, G.R. (2008). The 3rd ACR in Tal Afar: Challenges and adaptations. (January), Carlisle Barracks, Pennsylvania. *Strategic*

Studies Institute. Retrieved from www.strategicstudiesinstitute.army.mil/pdffiles/of -interest-9.pdf

McLoughlin, S. (2003, April 12). Rumsfeld on looting in Iraq: "Stuff happens". CNN.com /U.S., War in Iraq. Retrieved from www.cnn.com/2003/US/04/11/sprj.irq.pentagon/

Michael, K. (2018). Special operations forces as the "Silver Bullet": Strategic helplessness and weakened institutional and extra-institutional civilian control. In J.G. Turnley, K. Michael & E. Ben-Ari (Eds.), *Special operations forces in the 21st century: Perspectives from the social sciences* (pp. 59–73). Routledge.

O'Neill, B.E. (2005). *Insurgency & terrorism: From revolution to apocalypse* (2nd ed.). University of Nebraska Press.

Packer, G. (2006, April 10). The lesson of Tal Afar: Is it too late for the administration to change its course in Iraq? *The New Yorker*.

Ricks, T. (2006). *Fiasco: The American military adventure in Iraq, 2003–2005*. London: The Penguin Press.

Ricks, T. (2009). *The gamble: General David Petraeus and the American military adventure in Iraq, 2006–2008*. London: The Penguin Press.

Sageman, M. (2004). *Understanding terror networks*. University of Pennsylvania Press.

Sageman, M. (2008). *Leaderless jihad*. University of Pennsylvania Press.

Scott, W.J., McCone, D.R., & Mastroianni, G.R. (2009). The deployment experiences of Ft. Carson's soldiers in Iraq: Thinking about and training for full-spectrum warfare. *Armed Forces and Society, 35*(April), 460–476.

Segal, D.R., & Kestenbaum, M. (2002). Professional closure in the military labor market: A critique of pure cohesion. In D.M. Snider & G.L. Watkins (Eds.),*The future of the army profession* (pp. 441–452). New York, NY: McGraw-Hill.

Shaw, M. (2005). *The new western way of war*. Polity Press.

Totten, M.J. (2007, September 10). Anbar awakens, Part I: The Battle of Ramadi. *Michael J. Totten's Middle East Journal*. Retrieved from www.michaeltotten.com/archives /001514.html

Trinquier, R. (1985). *Modern warfare: A French view of counterinsurgency*. Ft. Leavenworth, Kansas: Combat Studies Institute, first published in France in 1961 under the title *La Guerre Moderne* by *Éditions de la Table Ronde*, Paris, translated into English by Daniel Lee and published in Great Britain in 1964 by Pall Mall Press, London.

Ullman, H.K., & Wade, J.P. (1996). *Shock and awe: Achieving rapid dominance*. National Defense University.

Warrick, J. (2015). *Black flags: The rise of ISIS*. Doubleday.

4   Bureaucracies, Networks and Warfare in a Fluid Operating Environment

Jessica Glicken Turnley

Introduction

The end of the Cold War, marked by the fall of the Berlin Wall in 1989, the horrific events of the first decade of the 21st century such as the 9/11/2001 attack on the United States and the Madrid and London subway bombings, and the political instability and conflicts in the Middle East, has prompted a huge literature about changes in the nature of warfare. The shift from a confrontation/conflict between two near-peer nation-states to asymmetric conflicts between a nation-state and non-state actors of various types has raised questions about the nature of warfare, many of which were addressed in the introduction to this volume.

Responses by nation-states to this new combat environment have included increased reliance on special operations forces (SOF), as demonstrated by significantly increased funding and manpower levels (Turnley, Ben-Ari & Michael, 2017). The argument is that these forces, with their small, highly agile combat units, are the nation-state's military resource which most directly mirrors the non-state actor as combatant and, hence, is best suited for operation in the new battlespace.

That being so, some militaries, including the US military, are teaching some of the critical combat skills of SOF to their "regular" military personnel (such as the Army's Security Force Assistance Brigade). This, then, begins to beg some larger operational questions. This chapter addresses one of those questions, as discussed in the introduction to this volume. Prompted by the changes in the nature and definition of the battlespace, what are some of the issues that might arise if militaries were organized – or perhaps just operated – by relatively transient operational missions? This shift in organization and operation would, it could be argued, better match the military to a highly fluid operational environment than does organization by more enduring criteria such as the traditional definitions of the battlespace (air force, navy, marines, army [ground force]). This discussion frames that question as one of the differences in the consequences of organizational form – more specifically, of the different consequences of hierarchic or bureaucratic and networked approaches to social organization and functioning.

Along with predictions of the end of the nation-state came predictions of the demise of its sovereign and bureaucratic structures of governance. Simultaneously, the first decade of the 2000s saw a parallel body of literature emerge heralding the rise of networked economies and societies connected by computerized networks and accompanying virtual marketplaces and social media platforms to replace those disappearing nation-states (see Castells, [1996] 2011, for one of the more influential texts). Discussions about formal organizational structures such as bureaucracies have become increasingly outpaced by discussions about network forms in the scholarly literature in recent decades. McEvily, Soda and Tortoriello (2014) tracked the appearance of the terms "social network" and "organization chart" (which they saw as a proxy for bureaucracies) in 5.2 million books published between 1937 and 2008 and digitized by Google and found that the term "social network" appeared more frequently than "organization chart" beginning in the late 1970s. The frequency of the term continued to increase at a rapid rate until approximately 2010 (the end date of their study), whereas "organization chart" began a slight decline in the late 1980s which continued until the end date of the study, when "social network" appeared about five times more frequently than did "organization chart".

Undeniably, studies of social networks have been enabled by technology to a great degree. Prior to the advent of computational power easily accessible by scholars, collecting data on networks was extremely laborious and time-consuming, and the statistical analyses required of the data ranged from being very difficult to impossible to perform by individual scholars. With the development of desktop computing came data manipulation and statistical packages easily available to scholars that enabled the development of sophisticated network analysis techniques (see, for example, Wasserman & Faust, 1994, and the voluminous literature produced by Kathleen Carley and her students at Carnegie Mellon).

This begs the question: were the networks always there and simply invisible to scholars because the tools to see them were not available, or have there been significant changes in social organization facilitated, to some degree, by these same computational tools? One thinks of Robinson Crusoe seeing a footprint on the beach after decades of believing himself alone on the desert island and his subsequent conclusion that "ignorance is bliss":

> It happened one day I was exceedingly surprised with the print of a man's naked foot on the shore. ... This renewed a contemplation which often had come into my thoughts in former times ... when first I began to see the dangers we run through in this life; how wonderfully we are delivered when we know nothing of it.
>
> (Defoe, [1719] 1968, p. 134).

The publication of such hugely influential texts such as Castells' ([1996] 2011) *The Rise of the Network Society* (over 41,000 citations in Google Scholar as of this writing) does suggest that the nature, not just the emphasis, of social networks has changed. Castells (1999) points out that,

> To be sure, networks have always existed in human organization. But only now have they become the most powerful form for organizing instrumentality, rather than expressiveness.
>
> (p. 6)

Castells (1999) argues that networks now *do* things (are instrumental), going beyond their historic, more affective role.

In this brave new world, the vertically structured, rules-based, top-down organizations known as bureaucracies have given way to what Castells ([1996] 2011) calls the "horizontal corporation" which is "a network of self-programmed, self-directed units based on decentralization" (p. 164). As Courpasson and Reed (2004) put it,

> If the 20th century … was the "age of organization" and bureaucracy was its core symbolic value and institutional mechanism, then the 21st century, contemporary social theorists and organizational analysts … tell us, will be the "age of networks' and the socio-technical infrastructure that make them indispensable as the new modes of collective cognition and governance required in revolutionary times.
>
> (p. 6)

However, some scholars decry what they see as an oversubscription to the virtues of networks. Davies and Spicer (2015) call it,

> network-boosterism; the tendency to attribute excessive analytical and normative weight to the role of networking in organizing and regulating relationships between governments, corporations and citizens.
>
> (p. 225)

Du Gay (2000), in his passionate defense of bureaucracy in the aptly named, *In Praise of Bureaucracy*, points out the potential loss in networked systems of bureaucratic values such as the professional pride generated by the meritocratic nature of the bureaucratic system. He also sees the disappearance in social networks of what Durkheim ([1938] 2013) called the superorganic or the *collective conscience*, and Courpasson and Reed (2004) called the "civic ethic". This recognition and investment in the whole stimulated the development of "core values such as duty, fidelity and pluralism [which] could flourish and infuse both the theory and practice of modern governance" (pp. 6–7).

Wilson (1989), in his detailed descriptions of government bureaucracies, comments that only through the continuity of bureaucratic forms can men be persuaded to act irrationally on a battlefield, giving their lives in pursuit of a rather diffuse end. Kallinikos (2004) points out that bureaucratic structures, invested with the professionalism that du Gay (2000) described, provided transparency, equality and the possibility for checks and balances that are missing in networks.

> Formal role systems provide transparent motives and legible behaviour, on the basis of which one can decide whether citizens are treated equally and the employees of the organization protected from any abuse or arbitrary exercise of power.
>
> (Kallinikos, 2004, p. 17).

This paper will begin with a short discussion of the nature of bureaucratic and networked organizational forms and their implications for the engagement of multiple types of players in a military context. It will then provide three examples of efforts in the US military to utilize networks to better leverage its own forces. Following the discussion of US military efforts, a US civilian management and coordination structure called the Incident Command System (ICS) will illustrate how *both* networks and hierarchic structures, always at play in all organizations, can be leveraged to effectively address fluid, high-tempo and potentially dangerous situations.

Networks and hierarchies

In a very general sense, organizations are organized and managed by two types of organizational structures – social networks and bureaucratically defined hierarchies. (For a much more detailed description of these two organizational types and their differing social functions than will be presented here, see Turnley, 2006.) Each organizational type is managed through a different mechanism of social control and addresses different types of social needs. All organizations exhibit both types of structures, although any given organization will emphasize one or the other.

Large organizations[1] usually are highly structured, enduring and bounded entities. They are organized for efficiency and perform very well under clearly defined and stable conditions. Activities, associated job functions and relationships among the jobs are defined by rules, processes, policies, laws and standard operating procedures. These descriptions also define functional relationships which are organized in hierarchies. These types of organizations are called bureaucracies (for detailed descriptions of bureaucracies, see e.g., Weber, [1947] 2013; Wilson, 1989).

Bureaucracies are traditional "line-and-box" organizations, where the "box" is defined by a social role which is described as a status and a set of associated and expected behaviors (e.g., "systems analyst" or "colonel"). Roles relate to each other through a set of explicitly defined processes and authorities, encoded in rules. Those rules generate and condone certain behaviors (e.g., military personnel salute those senior in rank to them; managers in an organization perform evaluations of their staff, not the other way around). Roles in an organization (note: not the people) are organized in hierarchical structures with top-down flows of authority. Because of the authority that is defined to attach to certain roles, the individuals occupying those roles can speak for more than themselves: they are authorized, by virtue of the social position they occupy, to speak for and to obligate the organization.[2] That same authority allows them to command certain individuals within the organization, but not others. (A colonel may command a lieutenant, but not a general.) Command relationships are asymmetrical. The colonel commands: the lieutenant obeys.

The totality of roles defines the organization, not the individuals who occupy the roles. The Weberian principle that roles in a bureaucracy are separated from the individuals who occupy them (Weber, [1947] 2013) makes individuals ephemeral in this organizational construct: they move in and out of roles and so (potentially) in and out of the organization as it exists over time. Thus, the depiction of the organization is atemporal: it could be of the organization at any point in time. The boxes in the organizational diagram are identified by roles, not names of individuals, and the totality of roles defines the organization.

In a bureaucracy, the organizational gaze is vertical: power inheres in roles at the top, and direction and instruction flow down. Responsibility is local to one's job. To refuse to take on an activity by saying "It's not my job" (i.e., is not included in the set of behaviors defined to adhere to this position) is to recognize this responsibility as local. Success is defined as a particular job well done, even if the organization as a whole fails to meet its goals. (The obvious exception is the individual at the top of the hierarchy. Success in his/her job is defined by the success of the whole organization.)

The second organizational type is the network. Social networks are dynamic, temporary configurations of relationships of individuals. Networks are composed of sets of dyads (where a dyad is two connected individuals) that coalesce around a purpose or a mission. When the mission is accomplished, the network dissolves. Networks are formed bottom-up by the elements that compose them. They are not formed top-down by a commanding authority.

Rather than a line-and-box depiction, as found in a bureaucracy, networks are represented by the now-familiar node-and-relationship construct.

Unlike a bureaucracy, in a node-and-relationship representation, the nodes are defined by proper names, not functions: they are actual persons. This locates the network in time and space in the way that the abstractions of the roles represented by a line-and-box relationship do not. The network exists, in part, because Ian and Sam, not Peter and Sam, connect. If Ian and Sam lose their connection, or Peter and Sam do connect directly, the network changes its character. Any depiction of a network references the network at a particular time.

In a network, the organizational gaze is horizontal to colleagues, not vertical to a superior or inferior. Relationships are consultative and advisory, not directive. That is, an individual exercises power only through his relationship with an Other. Relationships in a social network thus are based on trust in persons, and not on the power and authority derived from the position at work in a bureaucracy.

In a social network, relationships are formed because of some similarity between two individuals, a principle in network analysis known as homophily. The purpose or cause that stimulated the formation of the network determines the dimension of similarity that is relevant and defines the network. Individuals belonging to networks formed to address a social cause such as the protest against a regime, for example, will have in common antipathy toward that regime. An individual belonging to that network may also belong to a neighborhood network that coalesced around a leaky sewer. In the second network, the point of commonality among network members would be those affected by the sewer. Other members of one network may or may not belong to the second.

In a social network, commitment is global, to the enterprise. In fact, it is a commitment to the enterprise that defines the network and that can act as a strong recruiting mechanism. (For an example see McAdam, 1986, who studied recruits to the "high-risk/cost" cause of the American Freedom Summer of 1968 which involved registering Black voters in the American South, and his conclusion that commitment to the cause and some prior engagement with cause actors were two strong recruitment factors.) No one succeeds unless the enterprise succeeds. All participants are responsible for the whole, to solving the problem or resolving the issue.

Social networks are highly ephemeral. They coalesce around some problem or issue. Members are recruited (or volunteer) on the basis of their commitment to the network's central cause, among other factors. Once the problem or issue is solved or resolved, the network disappears, or the ties fade and become what are known as weak ties (Granovetter, 1973) – infrequently activated, but providing crucial connections when they are. As networks are formed in response to relatively transient issues, they seem to be an ideal organizational form for a fast-paced and changing world.

As all participate relatively equally in a network, no one has the authority to speak for the whole. As Stalder (2006) put it,

> Within the network, there is no single point that has the formal authority to impose its will on other network participants in the same way that decisions can be imposed by a chief executive thanks to her unique position within the organization.
>
> (p. 178)

In fact, in a network there is no concept of what Durkheim called the "superorganic" (Durkheim, [1938] 2013), a concept of the social whole that is greater than the sum of the individuals which compose a group and that exercises a somewhat coercive influence on the behavior of the individual. In a military context, this commitment to the whole plays out in many ways, often described in terms of cohesion (MacCoun & Hix, 2010). Many have argued that the affective or emotional ties created by membership in the type of enduring group that bureaucracy creates are critical for the cohesion that contributes to battlefield performance. This is an illustration of how the rational and cognitive investment in a whole that a bureaucratic organization provides allows space for the development of strong affective ties typical of a network. As Wilson (1989) puts it, "Soldiers fight when the men next to them expect to fight" (p. 46).

Networks composed of dyads created by organizations connecting, rather than individual persons, represent a special case in network analysis. This type of connection is of particular interest to this discussion because a "mission formation", as described in the introduction to this volume, is likely not to be individuals coming together to address a particular cause. Rather, it likely will be an element of one military (perhaps as small as a platoon) connecting with some other military element to address a common threat. In these cases, the participating organizations (the militaries) are hierarchic, for there must be some agency or voice in the network that allows it (the whole, the "superorganic") to make the connection.

This, then, is another example of the possibility of the coexistence of networked and hierarchic forms of interaction in the same organization. The discussion will return to this point later.

McSweeney (2006) gives a sampling of terms used to describe what many are calling a new epoch in social relationships:

> Whilst "post-bureaucratic" is one of the most widely used labels for the supposed new organizational type, other terms are also employed in the same sense, including: "post-hierarchical"; "the post-modern organization"; "the virtual corporation"; "the cellular form"; and the "boundaryless corporation".
>
> (p. 22)

There is much literature which takes the position that the shift in the geopolitical role of the nation-state, the rise in power of the transnational non-state actor and the development of globalized markets largely made possible through technology all require a strong focus on the power of networked organizations, a full shift in gaze away from bureaucracies. That said, there is not full agreement that we have now entered the "age of the network". Davies and Spicer, for example, have pointed out that the very trait essential to the formation of networked ties appears to be in decline:

> If levels of trust in society are low and declining, as Cook, Hardin and Levy (2007) suggested in a comprehensive study, it poses a challenge of logic to the notion that networks are proliferating and capable of fostering cohesion.
>
> (Davies & Spicer, 2015, p. 221)

On the other hand, the popular press, as well as many scholarly works, has praised the quick-response capabilities of networks, well suited for a world subject to a fast pace of change. Bureaucracies are notoriously slow to respond to crises. They generally respond to problems operationalizing rules by developing new rules, a process of ever-increasing elaboration that could be called "bureaucratic baroqueness".

Some argue that, as the pace of world events and processes becomes faster, the use of the individual as the organizing element in networks becomes a more efficient means of governance than bureaucracies. This is partially because networks eliminate what has been called the principal-agent problem and its associated tensions whereas bureaucracies create it and so must put in place mechanisms to mitigate it.

In brief, the principal-agent problem arises from the fact that, in a bureaucratic organization, the leader box – that is, the leader role, not an individual – establishes the end such as might be described in a commander's intent, or a strategic goal. The organization, guided by the manager box, provides the means. The leader (the principal) thus must take measures to ensure that the wishes of his office are discharged as he wishes, because his authority for executing them has been delegated to others (the agents) over whom he may not have full control.

This sets up what Braun and Guston (2003) have identified as four non-trivial problems for the principal:

- The problem of responsiveness – does the agent do what the principal wants?
- The problem of adverse selection – is the agent the best one for the job?
- The problem of moral hazard – will the agent do his/her best to execute at the highest level of quality? and
- The problem of decision-making and priority-setting: are the right problems addressed at the right times? (Braun & Guston, 2003, p. 310).

Bureaucracies set up systems of controls to help ensure that these principal-agent tensions are appropriately addressed. For example, the problem of responsiveness is addressed by putting in place a system of rewards and punishments to ensure that the agent does what the principal wants. The problem of adverse selection is highlighted in the US military by the development of the Security Assistance Brigades to take on certain aspects of military engagement previously discharged by SOF, suggesting that perhaps SOF are no longer the correct "agent" for those aspects.

Networks, on the other hand, are composed of individuals, not roles. The individual in a network never relinquishes his authority or his agency. Only individuals acting independently and cooperatively through the set of dyads that make up a network can make things happen. We are back to Castells's insistence on the instrumentality of networks and, thus, the importance of the individual.

Bureaucracies are often condemned for the proliferation of "red tape" that results from bureaucratic baroqueness. However, just as bureaucracies can develop pathological conditions, so can networks. Networks can be highly exclusionary, for example. Created on the basis of ties of similarity between individuals, those lacking that demographic or element are not only marginalized but become invisible. As an example of a network pathology in another dimension, recent problems Facebook has faced with data privacy/distribution illustrate the dark side of unquestioning faith in ties based on trust. As the tie is a voluntary one, there are no consequences for violating trust other than social disapprobation when the tie is breached.

Nation-state governments, themselves highly bureaucratic organizations, have tried to establish various forms of networked organizations and operations to further national ends. Note that these are conscious efforts to "create" networks using a top-down or command and control approach. They are not efforts to create spaces that allow the emergent, bottom-up approach that many see as the beauty of the network model to operate. In the latter case, individuals voluntarily associate and disassociate, creating unplanned patterns with interesting consequences. As a general comment on government-created "networks", Davies and Spicer (2015) observed that

Far from representing a radical departure from bureaucratic modes of governance, networks may often be a form of window dressing designed to make [bureaucratic roles and relationships] look more legitimate, attractive and up to date.

(p. 228)

Consider the effort of the attempted change in governance structures in the 1990s known as new public management (NPM), often discussed in the scholarly literature under the rubric of public choice. Governance structures based on public choice, or NPM programs, were efforts to govern by outcomes rather than processes (Osborne, 1993). NPM manifested in the US as an initiative called "Reinventing Government" promoted by then-Vice President Al Gore. Under NPM constructs, governments were to set up what looked like "networked" relationships with contractors to perform identified jobs. However, the Reinventing Government initiative did not prove to be durable. In fact, the "networks" were not formed around any cause other than the rather diffuse one of good governance. Relationships between contractors and government entities remained nothing more than market-based contracts.

The US military and social networks

The US military is, in many ways, a quintessential bureaucracy. A huge organization – perhaps the world's largest, with 1,333,351 active-duty personnel at the end of 2019 (www.dmdc.osd.mil/appj/dwp/dwp_reports.jsp, accessed January 2020) – it presents a complicated picture of interlocking hierarchies of civilian oversight (the Department of Defense), the military services organized by historical definitions of the battlespace (Army, Air Force, Navy, Marines, the new Space Command, and the somewhat anomalous Special Operations Command) and the tactical geographic commands (Central Command, Southern Command, etc.). This section will very briefly discuss an interservice mechanism of the US military known as the Joint Task Force (JTF), followed by a brief overview of an American program called the revolution in military affairs (RMA) which appeared in the 1990s, which was to "revolutionize" the military largely through the development and deployment of a network-centric environment.

The Joint Task Force

The US military has had mechanisms such as the JTF available to it for decades to organize in what appears on the surface to be "networked" forms. In theory, the JTF is a manifestation of a particular type of military governance at the operational level. It is designed to allow the military to form and reform itself around transient missions and tasks. Although the following discussion is about a particular organizational form – the JTF of the US military as codified in JP 0-2 – many of the principles it raises have applications for task forces in militaries of other nations. The following discussion will very briefly investigate the JTF as presented in US doctrine, showing it to be much closer to Davies and Spicer's (2015) window dressing than to a true network.

JTFs were (and still are) one of the US military's primary vehicles for operational, mission-focused engagement across military services, with the potential to capitalize on newly fielded communication and information-collection technologies. However, to this date, these JTFs are still created and operated through the doctrine that focuses heavily on the external and top-down creation of hierarchical organizations. JTFs are usually comprised of only military elements, although they can be structured to include other governmental elements and, if necessary, governmental elements from international collaborators. Note, however, that the JTF has no mechanism to include non-governmental organizations (NGOs) of any type.

There have been some notably successful JTFs in the US military. The Joint Interagency Task Force South, or JIATF-South, for example, has been able to successfully integrate the operational cultures of several organizations and multiple countries into a single cross-agency team focused on drug interdiction in the Caribbean (Munsing & Lamb, 2011). Task Force 714, led by General McChrystal in Afghanistan, was itself not a *joint* task force but was able to engage personnel from US intelligence and other US government agencies with SOF to significantly improve battlefield success rates (Schultz, 2016). The TF 714 effort will be described later in this section. There have been other task forces throughout the course of the recent Middle East conflict that have engaged participants from the international coalition with US forces in productive military interactions. However, led and structured by the US, these task forces in general were hierarchical, bureaucratically controlled organizations.

In principle, the formal US JTF sounds as if it is a mission-based force, designed to draw resources from across the military to address a specific problem. It appears to be ideally suited to an environment in which forces coalesce and dissolve as a wide variety of missions appear and disappear. In fact, as this discussion will show, it is a strongly hierarchic, bureaucratic form.

According to JP 0-2, the US military doctrine defining a JTF, a JTF is an operational-level force which may be established "on a geographical area or functional basis when the mission has a specific limited objective and does not require overall centralized control of logistics" (JP 0-2 2001, V-10). In order to be considered a JTF, the organization must contain elements from more than one service component. JTF commanders should "seek to minimize restrictive control measures and detailed instructions" (JP 0-2 2001, III-14) by communicating direction and purpose using the commander's intent and mission-type orders.

The commander's intent appears to speak explicitly to the network construct as it "represents a unifying idea that allows decentralized execution within centralized, overarching guidance" (JP 0-2 2001, III-14). Mission-type orders take the fluidity of the commander's intent (in theory) down to the task level. Mission-type orders

> direct a subordinate to perform a certain task without specifying how to accomplish it. The senior leaves the details of execution to the subordinate,

allowing the freedom – and the obligation – to take whatever steps are necessary to deal with the changing situation. This freedom of action encourages
the initiative needed to exploit the volatile nature of joint operations.

(JP 0-2 2001, III-14)

The construct behind the framing commander's intent and the delegation of execution to subordinates might appear to establish the independent nodes in a network. However, a delegation of authority to achieve a purpose recognized and
defined by a commander is qualitatively different than a purpose which forms the
catalyst for the engagement of two individuals into a network dyad. Rather than
a bottom-up formation with agency and authority residing in the lowest levels,
the JTF is a top-down formation, with agency delegated down, but with authority
remaining at the top. This sets up a classic principal-agent problem absent in a
true network, where all individuals are equally responsible for the execution of a
jointly defined intent.

The problems associated with organizing a JTF begin with the need to determine which commander will issue the orders and describe the intent. This is particularly contentious in a task force composed of elements from different services
with equivalent levels of command. There is no incentive, other than a commitment to a common goal, to relinquish command – and, through aspects of the
principal-agent problem, there are many disincentives to do so.

It is not surprising that an analysis of the operation and functioning of a JTF
by Olechnowicz (1999) argues that the most contentious domains within the
organization of a JTF are those of command and authority (pp. 47–50). In fact, JP
0-2 (2001) makes the centrality and the importance of command, its associated
authority *and the risks of delegation* very clear:

Command is central to all military action, and unity of command is central to unity of effort. Inherent in command is the authority that a military
commander lawfully exercises over subordinates and confers authority to
assign missions and to demand accountability for their attainment. Although
commanders may delegate authority to accomplish missions, they may not
absolve themselves of the responsibility for the attainment of these missions.
(III-1)

Note that, although the authority associated with the command may be delegated,
ultimate responsibility remains with the commander. The commander must have
mechanisms in place to ensure that the delegates perform as she/he wishes – a
clear incidence of the principal-agent problem described earlier.

As a result of the issues with command and authority identified by Olechnowicz
(1999) and others he himself raises, Dull's (2006) detailed analysis of the JTF's
suitability for what he calls "network-centered operations", argues that "The
stylized organizational architecture of a Hierarchy rises as the only applicable
architecture for a JTF" (p. 32). In fact, he goes on to say that, under current JTF
doctrine, the hierarchy so clearly segregates the service components that "Units

in the same relative echelon but under different components typically do not have the authority to coordinate and provide services for one another, much less task each other directly" (Dull, 2006, p. 62). Thus, despite the mission-type orders, the agency of JTF elements is severely restricted by bureaucratic rules. Underscoring the primacy of the bureaucratic structures, Dull (2006) also argues that "Informal communication may have the same effect at coordinating and sharing information as formal communications, but it is not a recognized relationship" (p. 55). The organizational gaze within the JTF is firmly vertical and strongly rooted in rules and procedures.

Finally, as noted earlier, there is no place in the JTF for guidance on including NGOs or civilian groups in the JTF, despite a strong statement on the importance of "interagency and civil interoperability" (JP 0-2, 2001, I-10). This is a significant lacuna in the doctrinal description.

In short, although on the surface a JTF might appear like an ideal place for a mission-driven military organization to emerge, Dull's (2006) analysis concludes that "Though the US military continues to implement the technology components of Network-Centric Operations, a corresponding evolution of organizational doctrine is not taking place" (p. 33). Dull (2006) goes on to say that, largely because of the conflicts in command and authority raised by the joint operation of units from different services,

> Current JTF command structures cannot attain the ideal of a flattened organization, and as described with regards to Army hierarchy, are likely the most hierarchical command structures possible
>
> (p. 58)

The revolution in military affairs/network-centric warfare

An appreciation of the value technology could provide to the implementation of network-based military organizations emerged in the US military in the 1990s. A recognition of the information-sharing value of new networked technologies and their potential to facilitate new organizational forms prompted Andrew Marshall of the Pentagon's Office of Net Assessment to proclaim a "revolution in military affairs" – a program later known as RMA. Although so-called RMAs are not new – historical examples include everything from the changes induced by the use of gunpowder to the blitzkrieg, made possible by the mechanization of warfare[3] – the RMA led by Andrew Marshall was intended to capitalize on emerging information, communication and warfighting technologies and to reorganize the military force accordingly. The RMA was billed as:

> a major change in the nature of warfare brought about by the innovative application of new technologies which, combined with dramatic changes in military doctrine and operational and organizational concepts, fundamentally alters the character and conduct of military operations.
>
> (McKitrick et al., 1998, p. 65)

The US RMA spawned a follow-on military concept of "net-centric warfare" which – at least conceptually – included the human and organizational component of the new way of war, as well as the technology (Alberts et al., 2000). According to a "functional concept" document published by the Defense Department in 2005 (United States Department of Defense, 2005[4]), a "net-centric environment" is

> a framework for full human and technical connectivity and interoperability that allows all DOD users and mission partners to share the information they need, when they need it, in a form they can understand and act on with confidence, and protects information from those who should not have it.
>
> (p. 1)

In this context, "mission partners" extends to many non-US military organizations, including,

> allies, coalition partners, international organizations, civilian government agencies, non-governmental agencies, and other non-adversaries who are involved with the activities or operations of the Joint Force.
>
> (p. 2, n. 6)

According to the functional concept, the technical capabilities associated with net-centric warfare would change the locus of integration among the military elements "by integrating the Joint Force across progressively lower echelons" (United States Department of Defense, 2005, p. 2). It thus would (presumably) push some decision authority down the ranks. This is supported by text later in the doctrine document, which reads,

> Each individual actor in the Net-Centric Environment has rights and responsibilities as they apply to information and decisions. This significant cultural shift must be supported by training and education. Individuals will have the proper incentives to fulfill their roles as producers, processors, and consumers of information. ... Individuals in the Net-Centric Environment also have decision rights and responsibilities and will be empowered and enabled to act freely in making decisions. They have the responsibility to make those decisions within the context of command intent and to share situation understanding across the Joint Force and its mission partners.
>
> (United States Department of Defense, 2005,
> p. 15)

The functional concept (United States Department of Defense, 2005) recognizes the difficulty of the human transition from a bureaucratic structure to a networked organization. It notes that the human transition

> requires surmounting internal and external organizational and policy barriers to the sharing of awareness, understanding, decision making, and the

synergistic application of force capabilities. This cultural change must be supported by training and education, as well as by ensuring that Joint Force elements have incentives to use the technical networks of the Joint Force and its mission partners to draw on appropriate capabilities, regardless of their geographic or organizational location.

(p. 11)

The functional concept (United States Department of Defense, 2005) outlines skills that military personnel must possess:

To operate successfully in this environment, people and organizations must be capable of dealing with flexible authority relationships (senior/subordinate, supported/supporting). This requires appropriate training, an understanding of the various organizational relationships, and the ability to work within an implied command intent environment.

(p. 22)

It then goes on to provide a laundry list of abilities necessary to work in a net-centric environment (adapted from United States Department of Defense, 2005, pp. 22–3):

- Ability to establish organizational relationships
- Ability to collaborate
- Ability to synchronize actions
- Ability to share situational awareness
- Ability to share situational understanding
- Ability to conduct collaborative decision making/planning
- Ability to achieve constructive interdependence

The last item on the list, the "ability to achieve constructive interdependence", is further defined as the ability to

employ ... the network (both human and technical) to allow a virtually limitless combination of latent Service and component capabilities in ways that create capabilities not previously achievable.

(United States Department of Defense, 2005, p. 23)

The exercise of these abilities appears to directly address the problem of fluid, rapidly changing missions, as outlined in the introduction to this volume.

However, despite the recognition of the importance of education and training to facilitate organizational change that the 2005 functional concept (United States Department of Defense, 2005) displayed, it also contains the following statement in its concluding section:

Within the Joint Force, organizational structures will transform as information and understanding are shared. New organizations will emerge, existing organizational structures will change (e.g., flatten), and some organizational structures will disappear.

(p. 31)

With this statement, the functional concept appears to imply that organizational change will "emerge" as a result of the deployment and use of information-sharing tools.

And, in fact, although the document lays out a blueprint for technology and organizational development, the human side lagged far behind the technical. As the US RMA unfolded, its focus on fielding innovative technology was accompanied by a blind spot on the organizational side. As the technologies were developed and deployed, parallel changes in organizational structure and the conduct of operations never materialized. As Dull (2006) points out,

> Technology appears to be outpacing organizational and doctrinal innovation, with commanders continuing to organize joint forces in doctrinal structures that have existed since before the first Gulf War.

(p. 10)

A much later thematic analysis by Tunnell (2015) of a range of Army materials[5] claimed that, although there was a relatively high level of interest in network-centric warfare concepts in the early 2000s, interest fell off quickly, and there was little to nothing written on implementation. In fact, as late as 2016, control over assets in the field still belonged to and was fiercely guarded by specific agencies and services, although the data and information they produced were service-agnostic (McNamara & Turnley, 2016).

Task Force 714[6]

TF 714 was a highly trained, elite US military counterterrorism organization that had been extremely successful with its historic missions of surgical strikes and hostage rescue. In 2003, it was deployed to Iraq, tasked to dismantle the al-Qaeda in Iraq (AQI) insurgency. The task force soon found that it was "losing to an enemy ... we should have dominated", according to retired US Army General Stanley McChrystal, then TF 714 commander (Schultz, 2016, p. 1). This realization prompted a redefinition of the task by General McChrystal and an accompanying reorganization of the task force and the way it operated to allow the task force to achieve its goal.

In Iraq, TF 714 faced an environment quite unlike the ones under which it had previously operated. In its early days in Iraq, the task force found itself successful at surgically focused and executed direct individual raids, carrying out about 18 raids a month (Schultz, 2016, p. 7). However, leadership found that this was not significantly affecting AQI activities; therefore, TF 714 was failing, unable to accomplish its task of dismantling AQI.

TF 714 leadership recognized that simply doing more of the same was not going to have the effect it was looking for. It realized that it had to make the organization more agile by raising the operational horizon beyond the individual raid. In order to do this, General McChrystal, TF 714's commander at the time, reconceptualized the task force's operational problem.

Historically, the task force had operated under a "find, fix and finish" (F3) approach. The task force conducted a raid, collected what materials it thought would be useful and detained appropriate individuals. Both materials and personnel were sent behind the operational lines to analysts from the Department of Defense (DOD) and other agencies who were trained in intelligence analysis and exploitation. Often weeks later, intelligence deemed useful by the exploitation and analytic organizations was returned to the operational front. Now, to become more agile, the task force needed to "find, fix, finish, exploit, analyze, disseminate" (F3EAD), that is, to perform many of the intelligence functions itself. For this to happen, General McChrystal needed to find the necessary expertise for the exploit, analyze and disseminate tasks he had added to the task force's job description. Expertise for these tasks was resident in the government, although it was not in his organization and mostly not even in his agency, the DOD. He had to bring it close to or inside his organization.

When TF 714 arrived in Iraq, it was organized and structured as a relatively traditional military organization, depicted by its leadership in classic line-and-block bureaucratic fashion. In his efforts to expand the nature of the task force's task, General McChrystal borrowed from the model developed and successfully deployed by JIATF-South.

General McChrystal engaged personnel from intelligence and other agencies in a cooperative model similar to JIATF-South. All involved in the F3EAD operations, regardless of their formal agency affiliation, were working for the same purpose or cause – the disruption of AQI through a series of targeted raids carried out by TF 714.

TF 714 achieved investment in the cause of all players through a reconceptualization of its operations that physically and organizationally co-located individuals from what were formerly discrete agencies. In facilities located near the point of operations (not on a base far behind the lines), operators and intelligence analysts now worked together in open bullpens that allowed immediate communication with anyone from any agency.

> The TF 714 headquarters was ... dominated by a large open space without cubicles, compartments, or limits on movement within it. It was an unrestricted, organic workspace whose aim was to facilitate the free flow of information and ideas through collaboration and interaction. ... At the front was a large U-shaped set of tables where General McChrystal, his component commanders, and senior interagency representatives sat.
>
> (Schultz, 2016, p. 47)

Open and quick communication among individuals regardless of rank, organization or job (military operational specialty) was a key part of the activities. As General McChrystal himself put it,

> any of them could walk freely across the room for quick face-to-face interaction. And with the touch of a button on a microphone, everyone's attention

could be captured simultaneously ... providing every component on the task force with an unobstructed, constantly up-to-date view of the rest of the organization.

(Schultz, 2016, p. 48)

Leaders of the raids were given more authority than they had under the previous model, moving decision-making closer to the organization's key operational behavior and locating it in those with the greatest behavioral knowledge and expertise. In this new model, leaders at what had been the "lower level" in the bureaucratic structure (leaders close to the raids but usually low in formal rank) became very important. They often were handpicked by the leadership rather than chosen through normal bureaucratic-military channels (Schultz, 2016, p. 60), or emerged through the demonstration of expertise.

In this way, TF 714 exhibited many qualities of a network. For example, all involved were highly committed to the defining purpose or cause, not just their "local cause" of intelligence analysis or raid planning and execution. All individuals had the potential for "equal say" when a question or action arose in their area of expertise. It also should be noted that, although there appeared to be a flattening of the organization, rank was not absent. The authority associated with rank was exerted if and when necessary. The rank of key individuals also continued to provide entrée for TF 714 to conversations at the strategic level.

It is important to point out that individuals were drawn into the "network" through their connection to General McChrystal. This is similar to McAdam's findings that one key factor for engagement of college students in Freedom Summer was some connection to those already involved. However, unlike the ICS model described in the following section, which has been formalized into a quasi-bureaucratic structure, TF 714's model cannot be replicated without the commitment and engagement of a charismatic individual with pre-existing ties throughout the national security establishment.

A civilian example: the Incident Command System

The civilian sector is also faced with crises of various types and durations. Although its crises may be different in some instances than the military's – natural disasters, for example[7] – the fluid, everchanging operational environment created by these crises is the same. In response, the US civilian sector has developed its own "interservice" mechanism to respond to this fluid environment. This mechanism, called the ICS, has evolved into an effective means to integrate governmental and non-governmental resources and forces to address both man-made crises (such as terrorist attacks) and natural disasters. It has done this through a novel combination of bureaucratic and networked organizational forms addressing what Choi and Brower (2006) called the "paradox" of emergency management: "The system must leave room for flexibility while it simultaneously provides accurate and timely communications and coordinates clear decision making" (p. 652).

The ICS is used by governmental elements at all levels in the U.S. (local, state and national). It provides a framework under which government and non-government resources, both formal and informal, can be effectively directed toward crisis mitigation. As such, it goes beyond the military JTF in several dimensions: given a focused mission, the ICS can incorporate both governmental and non-governmental resources; it recognizes mechanisms for bottom-up organization, and it uses the hierarchical command and authority structures to leverage network-based relationships during times of focused action. Its relatively structured and replicable framework also goes beyond TF 714's dependence on a single individual to activate and manage the interorganizational connections.

The ICS, as described by Bigley and Roberts (2001), combines some of the functional orientations of the bureaucratic structure with the reliance on individuals and their personal connections seen in social networks. The ICS has evolved from its beginnings as a response to problems associated with wildfire-fighting to an "all-risk" system suitable for almost any type of emergency that involves multiple tasks and agencies. It has been adopted by the Federal Emergency Management Agency as part of its formal response system (https://training.fema.gov/emiweb/is/icsresource/) and appears in the training programs of many emergency response organizations.

The ICS began as an informal coordination mechanism to fight wildfires in California in the 1970s as federal, state and local agencies worked to develop an effective coordination mechanism. "The critical innovation of the ICS was to temporarily centralize response authority to direct multiple organizations" (Moynihan, 2009, p. 896). This response authority was located in the incident manager. A crucial innovation of ICS is that the individual and/or the agency assuming the manager role will shift quickly according to the requirements of the incident at any given time.

An ICS is functionally organized, constructed around five major functions: command, planning, operations, logistics and finance/administration. Although this may initially look like an organization constructed along bureaucratic principles, there are some important differences. First, the organization is structured by functional requirements, not by units defined by functions. A second difference is one in which the ICS form, participant authorities, and activities are driven by immediate operational context rather than by predetermined and enduring roles and responsibilities. As Bigley and Roberts (2001) point out,

> The organizational approach must continually map to the requisite variety of a dynamic and risky situation. The system in use must be able to expand and contract, change strategic orientation, modify or switch tactics, and so forth, as an incident unfolds. Compounding the challenge are hazardous task contingencies that may not have been previously experienced or predicted. Furthermore, system and task complexities coupled with the need for immediate local adjustments may preclude the possibility of adequate or timely direction from superior hierarchical positions. Yet uncoordinated activities can drastically increase participants' hazard exposure. Finally, many or most

system members may have never worked together before and may have the realistic expectation of never working together again.

(p. 1286)

This is a complex, dynamic environment requiring a high degree of organizational flexibility with a strong focus on precise situational awareness and the response capabilities of the individuals closest to the event. Although the five functions identified earlier are all recognized by personnel who participate in ICSs, each function is staffed *only as the situation requires*. If personnel are "assigned to the boxes" although they are not needed, it takes resources away from places where they could be useful. The focus is on optimizing resources at the tactical level.

Each ICS is headed by an incident commander. The captain or head of the first unit on the scene assumes the role of the incident commander. He or she assesses the situation and asks for resources (such as local firefighters or humanitarian support) as needed. As those resources arrive, the incident commander assigns them as appropriate. At some point, as new personnel with higher rank arrive, the original incident commander may be reassigned, and the higher-ranking individual assumes the role. Unlike a command structure, such as the JTF, where leadership positions are defined a priori, the organizational location of the incident manager in an ICS changes throughout the course of the crisis as resource needs and other needs shift as the crisis evolves.

An effective ICS generally separates formal authority from decision-making. Although personnel do not give up formal rank positions when operating in an ICS (note how the original incident commander may get bumped based on rank), decision-making is pushed down the organization as required by the tempo of tactical operations. As the earlier quote from Bigley and Roberts (2001) noted, "System and task complexities coupled with the need for immediate local adjustments may preclude the possibility of adequate or timely direction from superior hierarchical positions" (p. 1286). Decision-making authority, in general, resides in those closest to the action. This allows low-level, expertise-driven teams to quickly form and dissolve, with their formation driven by conditions or problems rather than by formal organizational requirements. The teams are emergent, as networks are, and are defined by contextual requirements.

Bigley and Roberts (2001) found that, for an ICS to work well, all operating under the aegis of a single ICS must have a clearly defined and commonly accepted value system, often expressed through a common purpose. This common value system helps shape immediate goals related to the crisis and thus will shape the common understanding of the system. For example, life preservation and environmental protection are important values to all fighting a wildland fire. These common values will, in turn, develop a common, shared understanding of the operating environment during a fire or other crisis. Although Bigley and Roberts (2001) do not go far into this aspect of the ICS, requiring that all participants bring to the incident similar value systems goes deep into each of the participating organizations. It involves the development of organizational cultures, recruiting strategies, training and the like, and certainly involves those in leadership roles in more formal organizational settings. Clearly, the development of the value system that drives the purpose is work that has to take place prior to the incident.

Common value systems can drive the development of sensemaking and of common mental models (Weick, 1995). Bigley and Roberts (2001) found that all participants in an ICS bring to these complex and fluid task environments similar mental models of the event, which then constrain and drive action, although the level and type of detail will vary by the role and rank or authority of the individual. And, as the event unfolds, individuals of lower rank or authority will develop more detailed models of a narrower scope than will individuals of higher rank. Together, these "nested models" represent engagement with the total system. The mental models thus serve some of the same organizing and coordinating functions provided by the organizational structure in other cases, but allow for more systemic and organizational flexibility than would be found in a formally structured organization.

In some ways, the evolution of the ICS is captured in the urban and architectural design theories of Christopher Alexander. Assuming that users of a space know more about their needs than anyone else, Alexander developed an approach most memorably executed in the placement of paths in public spaces. Rather than have the architects design the path placements a priori, he reasoned, why not let the users of the space tell the architects where to place those paths by walking them (Alexander et al., 1975)? His designs for public spaces included no paths. After the spaces were built, over time, users of the space wore paths in the dirt as they moved from point to point. Those dirt paths were then formalized by pavement or some other mechanism. In the same way, the formal ICS was developed through practitioner experiments with various methods of organizing in real crisis situations. These means of organizing were captured and formalized in the ICS.

The ICS model is interesting for several reasons. It provides organizations with a reasonably defined and shared mental model for responding to relatively transient, highly dynamic and highly important events or conditions. It provides a shared vocabulary (all participants will know what is meant by a "wildland fire" or a particular class of emergency) and so gives a basis for the development of the required shared representations. It is able to manage devolved decision-making, with authority close to operations, while a pre-existing rank or authority structure continues to run in the background and can be activated if needed.

However, it is important to note that there is not universal endorsement of the value of a networked response to a crisis. The crisis response literature includes advocates of both a hierarchical command and control model and a network-oriented coordination and communication model which is based on collaborative processes (Moynihan, 2009, p. 897). Moynihan's (2009) analysis of several crisis case studies led him to emphasize the importance of centralized command and control.

The centralization imperative for crisis response is urgency. When combined with the need for a network of interdependent responders, urgency creates a coordination problem. Gradual processes of interorganizational consensus building and mutual adjustment take too long. Responders need a central coordinating mechanism to direct resources and resolve conflict in a timely fashion.

(Moynihan, 2009, p. 898)

Moynihan's (2009) analysis also led him to conclude that certain characteristics of networks as an organizational form make crisis response more difficult. These characteristics include the diversity of participating organizations and their associated values, which make it difficult to create common mental models, and the difficulties of sharing of authority among members, which is often subject to contention (as was highlighted earlier in the discussion of the JTF; Moynihan, 2009, p. 909). However, he also endorses certain aspects of a networked response and speaks critically of ICS shortcomings:

In particular, the ICS has been criticized for ignoring the importance of inter-organizational relationships, the spontaneous nature of response, the role of unorganized volunteers, and the potential for conflict between organizations.

(Moynihan, 2009, pp. 897–8)

He also notes the importance of enduring positive working relationships and trust, such as those General McChrystal leveraged for TF 714 (Moynihan, 2009, p. 909).

Provan and Kenis (2008) point up one of the very important factors that make networks a socially expensive means of organizing. They pointed out that "In networks, the primary tension regarding efficiency is between the need for administrative efficiency in network governance and the need for member involvement, through inclusive decision making" (Provan & Kenis, 2008, p. 14). Ensuring that all members of a network receive accurate communications and are equally involved in decision-making can be very time- and resource-expensive.

This tension between inclusiveness and efficiency can lead managed networks, such as those found in an ICS, to be relatively exclusionary when they are activated at a time of crisis. At that time of crisis there is no time available to work with other organizations to create common mental models based on shared values. One of the most important lessons for the military from the ICS is that inclusion activities can (and must) take place in the interstices of the crises. A relatively fluid and everchanging operational environment means that new situations are constantly emerging. However, few of these situations elevate to the level of crisis. It is to the benefit of organizations that are involved in crisis management to interact and engage when there is no crisis. As Moynihan (2009) put it,

The ICS illustrates the possibility of switching between centralized and more decentralized forms of network governance consistent with task demands. During crises, network governance is highly centralized. But between crises, the ICS does not exist. Crisis response networks are loosely affiliated and follow a shared governance model. In precrisis periods, responders can build working relationships and trust and improve their understanding of mutual capacities and the principles of the ICS, thereby laying the groundwork for an integrated response during the actual crisis. Crisis networks therefore suggest the fluidity of network governance.

(p. 911)

Note that, during inter-crisis periods, it is the people (the responders) who inter-act, not the agencies. As a respondent in one of Stark's (2014) case studies put it, "People are there to do that, not plans" (p. 702). Stark (2014) ultimately goes so far as to say, "agents promote adaptation within these constraints and that it is the innovation of agents, rather than the fabric of their bureaucratic environments, that allows efficiency and flexibility to coexist" (p. 704). General McChrystal's TF 714 worked because of General McChrystal's personal contacts. It is an illus-tration of how individuals engaged across services and agencies can activate in times of crisis personal connections they already had in place.

The importance of interorganizational trust during a crisis, even under a cen-tralized command and control structure, is interesting. Moynihan (2009) found that the inter-crisis periods when relationships could be built between organiza-tions and mutually intelligible communication structures created were essential to the effective functioning of a highly fluid command and control structure during crisis periods. Bureaucracies and networks can – and must – co-exist.

Authority, by itself, offers an inadequate basis for coordination without posi-tive working relationships between core members. ... This finding supports the claim that formal control systems and trust are not necessarily mutually exclusive alternatives to fostering coordination, but can complement one another, especially in high-risk contexts.

(Moynihan 2009, p. 912)

The size of the US military appears as if it would mitigate against the creation of these personalized relationships. The JTF, with its highly formalized rela-tionships of command and control, appears to be predicated on this assumption. However, while engaged with a project for the US Marines, the author witnessed the expression of such relationships. Traveling with a lieutenant colonel (of which there are about 2000 on active duty in the Marines) to all the main US Marine bases, with the exception of the base in Okinawa, Japan, the author wit-nessed her marine colleague activate pre-existing relationships with in-uniform colleagues at every base visited, relationships that greatly facilitated the execu-tion of the project.

Stark (2014) described the complementary roles of bureaucratic (hierarchic) and networked approaches as the "relevance of spontaneity and innovation as a means of enhancing adaptation but only within the pre-existing (rational) frame-work of crisis management" (p. 703). The National Incident Management System (NIMS) Training Development Guidance (under which ICS courses are offered) for the US Department of Agriculture says that "The NIMS is based on an appropriate balance of flexibility and standardization in order to provide a framework for interoperability and compatibility during incident operations" (www.usda.gov /sites/default/files/documents/NIMLesson.pdf).

Back to the beginning

The ICS – the civilian response to an unpredictable environment – although not perfect, promotes ways to engage networks of organizations, agencies and individuals in a fluid yet centralized command and control center. The incident commander is defined by the nature and stage of the crisis, not a priori as a role in an organization, as is a commander of a JTF. The position, with its authorities and responsibilities, thus has the possibility to be redefined by the players in real time as the situation evolves, forcing associated relationships to also redefine themselves. In short, the ICS structure illustrates the importance of McEvily et al.'s (2014) claim that

> both formal and informal organizational elements generate a web of inter-actions connecting actors. These interactions, whether formally designed or informally emergent, are conduits through which organizational actors coordinate efforts, share goals, exchange information, and access resources that affect behaviors and performance outcomes … [therefore] studying organizations from solely a formal or informal perspective is necessarily incomplete.
> McEvily et al., 2014, pp. 302–3)

McEvily et al. (2014) go on to assert that

> No organization exists, not even in the post-modern configuration, at either polar extreme of solely formally determined and prescribed behaviors or purely informally emergent action driven by individual agency. Instead, formal and informal elements co-exist in a variety of combinations and affect each other in important ways.
>
> (p. 302)

The interface between rules-based bureaucratic structures and relationship-based social networks presents many challenges to organizational leaders. Perhaps one of the most important points of tension is the tendency by those in authority in bureaucratic structures to "legislate" networks or "command" them to appear. This could take the form of dictating that certain individuals and organizations engage when a crisis or (in a military context) a mission demands it. Ideally, individuals at key contact points in organizations will develop relationships in the spaces between crises – relationships that then facilitate quick connections instigated by those closest to the action when situations demand. Equally inappropriate is the "requirement" to form a "cause-free" network (such as the "interagency" that many hoped would emerge in Washington, DC in the 2010s to combat terrorism) – to engage when there is no clear crisis or well-defined purpose to shape the relationships.

Another important lesson that can be learned from the examples here is that using the time between missions to establish relationships among personnel involved with various units (not among the units qua units) can be a very useful investment that can be realized in a time of crisis when mission-focused action becomes paramount. Using the networks representing the "paths" or relationships forged by those individuals, a command and control center can mobilize resources much more effectively than can the over-formalized JTF or a military force connected only by technologies. Creating those relationships while recognizing the value of certain aspects of bureaucratic standardization and the pitfalls of others can allow the creation of an organizational structure such as the ICS, which can leverage relationships and be transferrable from incident to incident in the way that General McChrystal's work with TF 714 was not.

This discussion has argued that bureaucratic, hierarchical and relatively stable organizations still serve some useful purposes, while mission- or purpose-driven networks have garnered more visibility in recent years and are able to be leveraged more effectively than they have been in the past. A military model similar to the civilian ICS could allow the military to retain formal structures as is required for a wide variety of military agendas while responding more quickly than bureaucratic structures generally allow to near-term and more transient situations. This model could provide a framework that would allow military resources to mobilize quickly around what may be an incident requiring international or coalition-based support extended to resources outside the governmental domain.

Engaging with a fluid, uncertain operational environment is very challenging. "Mission formations" in a military context – groups of individuals or organizations that coalesce around a particular task or purpose – are quite possible but not trivial. They require a recognition that the strongest organization is neither a bureaucracy nor a network: it is one which takes advantage of the strengths of each form and avoids their weaknesses.

Notes

1 Dunbar (1993) has argued that the human brain limits the number of relationships that can be effectively managed by one person to about 150, a size that has come to be known as "Dunbar's number". As groups grow larger than this rough number, they must default to forms that allow for one person to act as a proxy for a larger group, a feature of bureaucratic organizations. Recent studies (e.g. Gonçalves, Perra & Vespignani, 2011) have shown that Dunbar's number holds even in an electronically facilitated environment such as Twitter, although other studies (e.g. De Ruiter, Weston & Lyon, 2011) contest the validity of the number for human groups, suggesting that Dunbar's study of primates (from which he derived the principles that defined the number) did not constitute a legitimate analog for humans.

2 Much of the consternation over President Trump's 2018 announcement of the withdrawal of the US from the Joint Comprehensive Plan of Action (JCPOA), also known as the Iran nuclear deal, stemmed from exactly this consideration. President Obama entered into the treaty *on behalf of the United States*, not on behalf of himself. Although

Iran appeared to be fulfilling all of its obligations under the agreement, President Trump unilaterally withdrew the United States. That raised questions in the minds of many as to whether any treaty into which the United States entered was subject to the whims of the individual occupying the role responsible for engagement. Did, in fact, the president commit the country to honor the terms of the treaty, or did he only commit himself?

3 There is argument in the literature as to what actually constitutes a true revolution in military affairs. That argument is outside the scope of this discussion and is not treated here.

4 It is worth noting that, of this writing, there is no formal US military doctrine for net-centric warfare.

5 The Army is the largest US military service with 477,069 active-duty personnel at the end of 2019, compared with the Navy and the Air Force, the next largest services, which had 388,188 and 332,263 active-duty personnel, respectively (www.dmdc.osd.mil/appj/dwp/dwp_reports.jsp, accessed January 2020).

6 This example is taken from Shultz (2016).

7 Although the US military is increasingly involved in disaster response and other humanitarian missions.

References

Alberts, D.S., Garstka, J.J., & Stein, F.P. (2000). *Network centric warfare: Developing and leveraging information superiority*. Washington, DC: Assistant Secretary of Defense (C3I/Command Control Research Program).

Alexander, C., Silverstein, M., Angel, S., Ishìkawa, S., & Abrams, D. (1975). *The Oregon experiment* (Vol. 3). Center for Environmental Structures. New York: Oxford University Press.

Bigley, G.A., & Roberts, K.H. (2001). The incident command system: High-reliability organizing for complex and volatile task environments. *Academy of Management Journal, 44*(6), 1281–1299.

Braun, D., & Guston, D.H. (2003). Principal-agent theory and research policy: An introduction. *Science and Public Policy, 30*(5), 302–308.

Castells, M. ([1996] 2011). *The rise of the network society* (Vol. 12). Hoboken, NJ: John Wiley.

Castells, M. (1999). *Information technology, globalization and social development* (Vol. 114). Geneva, Switzerland: United Nations Research Institute for Social Development.

Choi, S.O., & Brower, R.S. (2006). When practice matters more than government plans: A network analysis of local emergency management. *Administration and Society, 37*(6), 651–678.

Courpasson, D., & Reed, M. (2004). Introduction: Bureaucracy in the age of enterprise. *Organization, 11*(1), 5–12.

Davies, J.S., & Spicer, A. (2015). Interrogating networks: Towards an agnostic perspective on governance research. *Environment and Planning. C: Government and policy, 33*(2), 223–238.

De Ruiter, J., Weston, G., & Lyon, S.M. (2011). Dunbar's number: Group size and brain physiology in humans reexamined. *American Anthropologist, 113*(4), 557–568.

Defoe, Daniel. ([1719] 1968). *Robinson Crusoe and the farther adventures of Robinson Crusoe*. New York: Washington Square Press. .

Du Gay, P. (2000). *In praise of bureaucracy: Weber – organization – ethics*. Newbury Park, CA: Sage.

Dull, D.S. (2006). *Implementing network-centric operations in joint task forces: Changes in joint doctrine* (No. ATZL-SWD-GD). Fort Leavenworth, KS: Army Command and General Staff College.

Dunbar, R.I. (1993). Coevolution of neocortical size, group size and language in humans. *Behavioral and Brain Sciences, 16*(4), 681–694.

Durkheim, E. ([1938] 2013). *Durkheim: The rules of sociological method: and selected texts on sociology and its method.* London: Palgrave Macmillan.

Gonçalves, B., Perra, N., & Vespignani, A. (2011). Modeling users' activity on Twitter networks: Validation of Dunbar's number. *PLoS One, 6*(8), e22656.

Granovetter, M. (1973). The strength of weak ties. *American Journal of Sociology, 78*(6), 1360–1380.

Joint Publication 0-2 (JP 0-2), Unified Action Armed Forces [UNAAF], 10 July. (2001). United States Joint Chiefs of Staff.

Kallinikos, J. (2004). The social foundations of the bureaucratic order. *Organization, 11*(1), 13–36.

MacCoun, R.J., & Hix, W.M. (2010). Unit cohesion and military performance. In B.D. Rostker, S.D. Hosek, J.D. Winkler, B.J. Asch, S.M. Asch, C. Baxter, N. Bensahel, S.H. Berry, R.A. Brown, L. Werber, & R.L. Collins (Eds.), *Sexual orientation and US military personnel policy: An update of Rand's 1993 study.* Santa, Monica, CA: RAND.

McAdam, D. (1986). Recruitment to high-risk activism: The case of freedom summer. *American Journal of Sociology, 92*(1), 64–90.

McEvily, B., Soda, G., & Tortoriello, M. (2014). More formally: Rediscovering the missing link between formal organization and informal social structure. *The Academy of Management Annals, 8*(1), 299–345.

McKitrick, J., Blackwell, J., Littlepage, F., Kraus, G., Blanchfield, R., & Hill, D. (1998). The revolution in military affairs. In B. Schneider & E. Grinter (Eds.), *The battlefield of the future – 21st century warfare issues.* Air War College, Maxwell AFB, AL.

McNamara, L., & Turnley, J. (2016). *An ethnographic study of culture and collaborative technology in the intelligence community.* Albuquerque, NM: Sandia National Laboratories (Distribution restricted).

McSweeney, B. (2006). Are we living in a post-bureaucratic epoch? *Journal of Organizational Change Management, 19*(1), 22–37.

Moynihan, D.P. (2009). The network governance of crisis response: Case studies of incident command systems. *Journal of Public Administration Research and Theory, 19*(4), 895–915.

Munsing, E., & Lamb, C.J. (2011). *Joint Interagency Task Force–South: The best known, least understood interagency success* (Vol. 5). Washington, DC: National Defense University Press.

Olechnowicz, S.M. (1999). *Identification and evaluation of organizational structures and measures for analysis of joint task forces* (Master's thesis). Monterey, CA: Naval Postgraduate School, December.

Osborne, David (1993). Public productivity & management review. *Fiscal Pressures and Productive Solutions: Proceedings of the Fifth National Public Sector Productivity Conference* (Summer, 1993), *16*(4), 349–356.

Provan, K.G., & Kenis, P. (2008). Modes of network governance: Structure, management, and effectiveness. *Journal of Public Administration Research and Theory, 18*(2), 229–252.

Shultz, R. (2016). *Military innovation in war: It takes a learning organization* (JSOU Report 16-6). MacDill AFB Press, FL.

Stalder, F. (2006). *Manuel Castells: The theory of the network society*. Cambridge, UK: Polity.

Stark, A. (2014). Bureaucratic values and resilience: An exploration of crisis management adaptation. *Public Administration, 92*(3), 692–706.

Tunnell, H.D. (2015, October). The US Army and network-centric warfare: A thematic analysis of the literature. In *MILCOM 2015–2015 IEEE Military Communications Conference* (pp. 889–894). New York: IEEE.

Turnley, J.G. (2006). *Implications for network-centric warfare* (No. JSOU-R-06-3). Hurlburt Field, FL: Joint Special Operations University.

Turnley, J.G., Ben-Ari, E., & Michael, K. (2017). Special operations forces and social science. In J. Turnley, K. Michael & E. Ben-Ari (Eds.), *Special operations forces in the 21st century: Perspectives from the social sciences* (pp. 1–12). Oxfordshire, UK: Routledge.

United States Department of Defense. (2005). *Net-centric environment, joint functional concept 1.0 7 April 2005*, http://extras.sltrib.com/Utah_Data_Center/netcentric_jfc-1.pdf (accessed January 2020).

Wasserman, S., & Faust, K. (1994). *Social network analysis: Methods and applications* (Vol. 8). Cambridge, UK: Cambridge University Press.

Weber, M. ([1947] 2013). *From Max Weber: Essays in sociology*. Oxfordshire, UK: Routledge.

Weick, K.E. (1995). *Sensemaking in organizations* (Vol. 3). Newbury Park, CA: Sage.

Wilson, J.Q. (1989). *Bureaucracy: What government agencies do and why they do it*. New York: Basic Books.

5 From leading combat units to leading combat formations

Modularity, loose systems and temporariness

Eyal Ben-Ari[1]

Introduction

What does command look like in mission formations, those temporary, often ad hoc, military configurations established for a specific mission (Introduction, this volume)? In this chapter, I chart out the main characteristics of such command at the tactical level. My work follows King's (2019) masterful analysis of contemporary divisions; he suggests that, in the 21st century, the command includes the executive role of making decisions, the managerial function of coordinating components to achieve the most effective results and the leadership part, centering on creating a common purpose and motivation. As my empirical exemplification I will use two documentary volumes published by the Swedish Centre for Studies of Armed Forces and Society (Tillberg & Tillberg, 2013; Tillberg et al., 2017). The two books consist of excerpts from interviews and conversations with officers and troops from the Swedish armed forces who had been posted abroad around the world since the 1990s. I use these sources both because they are rich in information and insights and because the missions that Sweden participated in fall squarely within the "New" or "Hybrid" Wars of the 21st century (Hoffman, 2007; Kaldor, 2001; Munkler, 2005). To be clear, however, I use these qualitative data sets to think through and illustrate my wider argument about tactical military command today.

As I read the two volumes, *Mission Commander* (Tillberg & Tillberg, 2013) and *Mission Afghanistan* (Tillberg et al., 2017), I was struck not only by the fascinating accounts but also by the extended journeys of self-discovery that many of the interviewees underwent. As is summarized wonderfully in the analytical introductions to the volumes, these journeys were characterized by trial and error, improvisations in contradictory contexts, meeting unforeseen challenges and navigating uncertain and unknown terrains. Indeed, the editors of the two volumes have developed many of these insights to offer readers an absorbing analytical framework for understanding the emergence and possible application of the actionable knowledge embedded in the interviewees' stories (Schon, 1983). But, upon reading the two books a second time, I soon realized that the interviews strongly resonated with wider issues that the social scientific study of the military has been engaged with during the past few decades. Accordingly, in what follows,

I suggest some thoughts about hopefully interesting aspects of command and leadership in today's conflicts that emerge from the accounts. To reiterate then, I utilize the rich material found in the volumes to rethink some of the accepted givens of the study of combat units and of tactical leadership today. Let me explain.

Although the "classic" combat units prescribed in military handbooks and textbooks – that is, platoons, companies or battalions – continue to be of crucial importance in all of the missions related to contemporary conflicts, and although many of the "model" traits of tactical leaders continue during such assignments, there seems to be something else that is *also* involved. As I read the volumes, I constantly encountered fascinating examples that did not fit these ideas. Missions such as those centered on mentoring, teaching across national military boundaries, short-lived task forces or multinational amalgamations involved the creation of new, temporary organizational forms that, although different, of course, from textbook units, are nevertheless characterized by their own structure, internal dynamic and indeed leadership styles (see the Introduction to this volume). Indeed, it seems that, beyond platoons and at times companies, there is very little appearance of stand-alone textbook units in today's armed conflicts. Numerous other examples of such temporary, ad hoc military forms – I shall call them mission or combat formations – can be found elsewhere in the two volumes in such instances as modular forms organized for high-intensity policing, provincial reconstruction teams (PRTs), ensembles for humanitarian work or staff work in UN missions. Although examples of such forms can no doubt be found historically, what seems to have happened is that the circumstances of the New Wars have constrained and obliged the Swedish military – as all of the participating armed forces – to use such arrangements much more than before. To use military parlance, the move in many current missions is toward designing and creating special mission-specific flexible-composition task forces and not fixed organic readiness units.

Along these lines, I suggest thinking about current-day military leadership not only in terms of the classic framework of textbook units but also in the context of these new combat formations. In this sense, I go beyond an emphasis on jointness as that term still assumes relatively homogeneous, bounded units (the textbook units) that are joined to other such units. Rather, I would emphasize the integration of diverse elements in a dynamic manner that may constantly change. Thus, for our purposes, an infantry company calling in air support or an Afghan–Swedish force on an operation becomes such a combat formation. In this chapter, I build on studies I carried out with the late Boas Shamir (Shamir/Ben-Ari, 1999, 2000, 2009), work with Uzi Ben-Shalom and Thomas Brond (Introduction, this volume) and previous work (Biehl, 2008; Bury, 2019; de Ward & Kramer, 2010; Hasselbladh & Yden, 2020; Schilling, 2019; Shields & Travis, 2017) to offer a framework for understanding the challenges facing military leaders in tactical-level combat formations.

Four formations, multiple issues

Let me trace out some types of formations that are found in the books because they illuminate the analytical points I am making, provide a basis for the rest of the investigation and can be found among all of the militaries of the industrial democracies.

Momentary leadership

First, take an instance of the most basic military relationship, a temporary, dyadic one linking leader and follower. Liridona Dauti (MA, p. 113), a captain (and company intelligence and liaison officer) reports about an operation, led by the Afghan National Police and in which a Swedish element was included, that came under shelling fire. Mastering her emotions and using her body language, she provides an Afghan policeman with confidence and direction, driving him to move. Although not under her command, he nevertheless looks to her for leadership and as a model for behavior. The two individuals are, for this very short duration, a social and organizational unit.

Leaders and interpreters

Per Thornell (MA, p. 103), an infantry platoon commander in Afghanistan, relates an episode involving his unit's interpreter. The latter is suspected of cooperating with the enemy in directing fire against the Swedes during a combat mission. As a result of his suspicion, Thornell confiscates the interpreter's cell phone so that he cannot contact the network with which he is perhaps cooperating. Yet, sociologically, what is the place of such interpreters? They are outside yet somehow inside the military; they are part of but also exempt from the hierarchy; they can be depended on but only to a degree. Are they led? Consulted? Given orders? Organizationally, they are part of temporary entities established for the duration of deployment or operation – an organizational form produced for a certain set of tasks.

Mediation

Mona Westerlund-Lindberg describes her experiences in leading a group of four women – each with her own field of expertise: gender advisor, military interpreter, doctor and herself a chaplain – to a female *shura* aimed at strengthening the position of women in a community and at gathering information that may interest the International Security Force that the Swedish contingent belongs to (ISAF). These four women are collectively a temporary modular unit for a short period that carries out the role of boundary-spanners between the military and local civilians. In this sense, they are sort of extensions of the military into civilian society and an extension of civilian society into the military. The advantage of such forms lies in their ability to perceive the needs and views of civilian groups and render them into concrete suggestions that commanders and troops can take into consideration. In this sense, this ad hoc formation is between and betwixt the military and its environment; hence, it is also part of practices that are not fully military or fully civilian.

Mentoring and combat formations

Third, take the instances of mentoring between individuals and between units as reported on by Martin Liander (MA, p. 193) and Bo Rahmstrom (MA, p. 131). Here, I suggest a shift of attention away from the complex of (often mutual) teaching/learning that goes on in such frameworks to look at the organizational forms through which these take place. Liander (MA, p. 193), a lieutenant colonel, commands a Swedish operational mentoring and liaison team to support the Afghan National Army (ANA). It is tasked to provide training, support, advice and actual cooperation with the local unit. The arrangement is for Swedish military personnel to spend, by and large, all of their waking hours with the ANA section they have been assigned for mentoring. Before deployment, Liander recruits and handpicks the participants for this temporary formation, which then begins to jell during preparations and planning. The 44 members of this "unit" are divided into mentors and support personnel (protection sections, medical unit, logistics and radio operators). Liander decides that the soldiers, and not only officers, will also be trainers and mentors to the Afghan soldiers, thereby obtaining twice as many teaching staff. But this turns out to be a problem as, in Sweden, soldiers are not allowed to supervise live-fire training; hence, it takes him "ages to have the order changed" by explaining the unique situation on the ground to the commanders in Stockholm. In addition, 12 interpreters are attached to this military formation, despite the fact they actually have no armed training. The mentoring team, in turn, is designed to be autonomous – jokingly referred to as "Travel light, freeze at night" – and soon finds that the "normal" purported rhythm of 10 days in the field and 2 to replenish often varies. Moreover, after devising a training and liaising plan, Liander and his staff meet with resistance from the rest of the Swedish (national) contingent (to Afghanistan) owing to different perspectives and "Complicated chain-of-command relationships combined with our SOPs differing from those of other Swedes in many respects".

In regard to ties with the Afghans, Liander observes that the "relationship is, of course, based on sharing the same hardships, and they are not always pleasant". No less important, the assumption shared between the two sides is that, if the Swedish mentors plan with the Afghans but are not there in execution, then "you are back to square one and they won't want us there. We also noted that the more difficult the situation we ended up together, the greater the confidence the Afghans had in us" (p. 196). But things are more complex, as a mentoring relationship – within an ad hoc formation – entails different dynamics that those found in textbook combat. Although the Swedes participate in planning and implementation, they do not decide what will be carried out but are dependent on the Afghan battalion. Yet this is not a one-sided dependence because "We were also to support their missions by putting ISAF resources their way. This was called "partnering". Quite often partnership enabled us to provide advanced resources, such as air support and mine-clearing capabilities for the Afghan plans". Accordingly, the partnering between the Swedes and Afghans is very different from the clear hierarchies described in textbooks.

In addition, and unfortunately, as Liander points out, this is a limited partnership, as ISAF could not always provide the resources owing to national restrictions on participation in areas with a high threat – usually, the areas where the Afghans were operating. As he well realizes, the Afghan battalion commander "wanted us with him because he then knew that with us in tow he was getting more combat units, planes and useful things. It was like a *symbiotic relationship* between them and us". And Bo Rahmstrom (MA, p. 131; part of another Swedish operational mentoring and liaison team), mentoring Mustapha, an Afghan battalion commander with much more experience than him, observed, "He is better than me in most things" as he has combat experience, language ability and knows the history and social codes. At times, Bo feels that Mustapha is mentoring him. And again, as in Liander's case, there is a theme of reciprocity and joint experience through which the relationship between Swede and Afghan emerges: "I feel as if I am letting Mustapha down by taking part in the planning but not in the execution".

But there is yet another layer of complexity here. Liander finds another split pitting the Swedish mentors against the Afghan battalion rather than the Swedish PRT members. This is a divide between the different realities of the unit and of the contingent that is expressed in practice by the relative unwillingness of PRT members to take on risks. "The consequences of that was that my soldiers rolled out along with the Afghans, but the PRT soldiers did not. These factors definitely did not strengthen relations between my soldiers and the PRT personnel." The positive side of this situation was the gradual emergence of trust and cohesion within the ad hoc formation between some of the Swedes and the Afghans:

> I think that the fact that we were always with them meant that they saw us in a different light [than PRT personnel]. They knew that invariably we went with them, and gradually as we conducted joint operations, they knew that we stood alongside them even when things were at their worst. I assume things happen to people when you share the experience of over thirty firefights together.

Looked at as a whole, the mentoring entity – a combat formation comprising Swedes, Afghans and occasionally other units such as air support or mineclearing – is a rather peculiar kind of military form that is far from the purported prescriptions of military textbooks. To be sure, it possesses a formal structure and division of labor; it is marked by ranks and procedures and has goals and objectives. But it is also something that is not only negotiated between individuals and dependent on the goodwill of the participants for compliance but is also characterized by an essential ambiguity between the two main elements – the Afghan and the Swedish – an ambiguity centered on mutual dependence, unequal resources and the ineffable quality of trust. At the same time, however, this is the kind of formation that features prominently in some current-day conflicts.

High-intensity policing

The final example of a type of formation is one reported on by Hakansson (MC, p. 37), who is a deputy battalion commander in Kosovo who suddenly finds himself in a very tense situation as battalion commander when his superior goes on leave ("I had to change roles very quickly"). He is at a disadvantage as he does not know the Swedish company commanders as well as his superior does and thus is not sure how they will react to tense situations. He illustrates this point by saying that he does not know them well enough to tell by their tone of voice if things are getting serious. As he takes up his role as commander and goes out of the base camp, he receives a report from Pristina that hundreds of Albanians are attacking Serbs and that what seems to be emerging is a very serious situation.

The company commander on the spot is dependent on the local police to control the situation, but just in case has arrayed his troops behind the police officers (fully equipped and looking, in his words, like "robocops"). It is at this point that a continuously growing organizational entity begins to form – a combat formation centered on high-intensity policing. The Swedish company drives through the village and joins the Finnish–Irish reinforcements (one company of each) sent to them, thereby becoming a multinational force that actually now comprises four companies (another Swedish company under Hakansson's command is alerted) and his small command post. Quickly following, special riot police comprising between 200 and 300 officers from Jordan, Pakistan, Bangladesh and Ukraine (and commanded by a French Gendarmerie lieutenant colonel) are added to the emerging force. And, to complicate things even more, a fifth Czech–Slovakian company is then added. This company, however, is not trained and equipped for riot control and has a national caveat stipulating they must not be used in such missions. It is thus given the task of patrolling the perimeter of the area. Somewhat later, an order to the Czech company commander to defend a monastery turns out to be difficult because of language problems, and Hakansson has to repeat the order a number of times to make sure that he is understood.

Back in the village, as violence escalates, police officers not belonging to the special riot units arrive to help. Hakansson finds himself in complicated circumstances, being in charge of and responsible and accountable for the formation that has been created but, at the same time, also dependent on the cooperation of other partners in an escalating situation. Cooperation with a French Gendarmerie lieutenant colonel, for example, entails working together across different "logics" of action: a military one based on maneuver and protecting flanks and a police one emphasizing maintaining a straight line in the face of protesters. Communication takes place both in face-to-face meetings and via radio and cell phones.

Another problem of communication emerges when a US helicopter unit is seconded to Hakansson as reconnaissance support. The difficulty is not one of language but rather of linking the languages of different service languages. As it turns out, an administrative officer who has asked to patrol with him that day becomes his "forward air-controller". Although not formally a controller, this officer has had past experience with the US Air Force and can communicate with its members by talking "aviation language". After a short while, yet another company joins the emerging force, a Norwegian one that Hakansson hands over to one of the Swedish company commanders on the ground as he is charged with overall command and control. To make things even more complicated, the Swedish special forces contingent stationed with them at base camp – who "did their own things", with Hakansson's battalion only providing "Food and potatoes" – offers to help. The special forces arrive in about a dozen jeeps and proactively ask him what he wants them to do. He asks them to put a stop to the Serbs' firing. The final element that Hakansson remarks about is his relationship with the media. After things calm down, he notes that, whereas the Swedish media are friendly, CNN and the BBC, who do not care about whether the mission was carried out successfully, look at him as the representative of all of KFOR (Kosovo Force) and want to know why KFOR has not succeeded in protecting the Serbian population.

All in all, Hakansson finds himself in charge of troops and police from at least ten different countries (with different national stipulations about the use of force and the tasks they can carry out), six companies (with leaders marked by different linguistic capacities in English), a contingent of riot police and some regular police officers (marked by their own logics of action), an American helicopter unit (that has to be communicated with using "aviation language") and some Swedish special forces (marked by relatively high autonomy in carrying out their missions). Were this not enough, this emergent combat formation is temporary, can change instantly (units can be taken away or linked to it) and operates within an explosive situation marked by protest, shooting and looting.

Although not as detailed as this case, other accounts in the volumes provide other fascinating examples of such created formations. Jan-Gunnar Isberg (MC, p. 139), deputy force commander and brigade commander in the Congo, notes that dealing with complexity in exercises in Sweden greatly aided him in Africa. He finds that this experience is relevant when he decides to launch a strike to disarm a local militia. He organizes a task group with the battalions under his command contributing units to the assembled formation. Isberg commands them from a command helicopter, and with him are senior officers from the different contingents through whom he relays orders. "When we were finished, the units returned to their normal deployment positions." In a like manner, Bo Brannstrom (MC, p. 53), then a brigade commander, is charged with more than 2000 soldiers from eight different countries. He finds himself constantly juggling resources and the different experiences of the battalions. For example, in contrast to the Swedes, some other countries in the KFOR force did not have the in-theater experience of taking a company from one battalion and sending it elsewhere to be under someone else's command. This kind of skill, he notes, one of giving and receiving units, must be learned. And Mats Strom (MC, p. 89), a deputy brigade commander in KFOR, was in charge of a staff that was itself an ad hoc combination of 60 personnel comprising professional military and reservists, trained civilians, women and men and Swedes and Danes. All of these combat formations are rather typical of the kinds of formations with which today's armed struggles are waged.

Combat formations, organizational fusions
and ad hoc amalgamations

Clearly, these types of missions and tasks and the organizational forms are intentionally or spontaneously created in ways allowing military forces to adapt to contemporary conflicts. As Cronin (2008, p. 1) elaborates, these conflicts are all profoundly political, intensely local and protracted. Indeed, the volumes are rich with examples of the attributes of these conflicts such as "fuzzy" categories of locals, shifting communal alliances or of the political implications of military action. Par Naslund, a warrant officer (MA, p. 106), recounts feeling exposed as one does not clearly know who the enemies are. Major Bo Rahmstrom (MA, p. 115), commander and mentor to an Afghan battalion, describes a scene when his unit reaches a valley in which heavy fighting has been taking place and there are dead lying on the ground. He then asks himself, "What if we get accused of this? How do I prove it wasn't us?". Clearly, he is aware of the international repercussions of this mission. Mattias Otterstrom (MA, p. 113), an infantry company commander, comments on the restraint needed in such conflicts, and Hans Ilis-Alm (MC, p. 133) observes that, in the Congo, he was not only a soldier but a diplomat and politician.

Against this background, what I focus on here are the organizational forms – the combat formations rather than "units" – that are created in such circumstances. Analytically, the concept of combat formation is intended to capture the (ongoing) processual nature of the various amalgamations, assemblages or combinations that military involvement in today's conflicts necessitates. As we saw through the various examples, at the tactical level (up to the brigade), they take place across organizational boundaries in ad hoc, specially created forces that *always* include a potential for the use of organized state violence. This potential cannot be overemphasized as, at the core of all of these missions, there are combat troops and combat units who specialize in the use of military force – the veritable expertise of the armed forces.

As is evident from the cases I have described, these various combat formations may take any combination of multinational guises, with armed units joining *gendarmeries*, police and international and local civilians. Some may last as little as a few hours or a day, some as long as a week or month and some for a few years (think of the PRTs). Yet, despite being impermanent and involving multiple actors within and outside the military, they are characterized by common goals (albeit at times limited ones) and entail constant interactions and negotiations between members. Of no less importance, such combat formations often contain multiple lines of authority (national and force), hybrid roles (soldier-diplomat, soldier-humanitarian worker, soldier-police officer), constant transitions (between roles, situations, organizational configurations), multilingual and multicultural communication and on-the-job training. Accordingly, military organizations, like all organizations, have become flexible organizations characterized by the relative loosening of internal and external boundaries. These "boundaryless organizations" (Davis, 1995), to stretch one image, contain fewer fixed structures and more temporary systems, whose elements, both people and technologies, are assembled and disassembled according to the shifting needs of specific projects. Yet the problem in all such formations is that they entail cooperating with entities marked by different interests and priorities, modes of communication and thought or practices and internal agendas.

In addition, these formations are often loose, temporary structures, sometimes marked by the unclear division of labor and authority, and are political arenas in which constituent actors promote and advance their own ends (Winslow, 2002). This point comes out most strongly in the cases of mentoring and in working with the PRTs. The mentoring formation, as is clear in the depiction by Liander, is full of tensions about mutual dependence, the authority to issue orders, and cooperation in and around planning and implementation. In PRTs, tension may arise between military officers and the civilians who deal with political and developmental tasks and the police who handle law and order. But tensions are also related to the military personnel within the PRTs who may become "civilianized" in a sense not only of taking orders from members of other forces or from civilians not of their own nationality, but of accepting greater limitations on taking risks than those in the field. In such situations, military leaders may find themselves not only in leadership roles but also in peer or subordinate roles (de Ward & Kramer, 2010). So the broader question is, what does the role of leaders in these various combat formations entail?

And military leadership?

To begin, tactical military leaders continue to need many of the characteristics and abilities that have long been necessary at this level: for instance, having at least the minimal psychological and physical levels of their subordinates, along with initiative, problem-solving, a proactive attitude and courage. There are many examples of these traits sprinkled throughout the volumes. Liridona Dauti (MA, p. 113), mentioned before, exemplifies the importance of military leaders as role models, especially under fire. In a remarkably candid account, Sergeant Henrik Nestow (MA, p. 105) relates how, in combat, he learned to overcome paralysis and surreal feeling to gradually return to a sense of control (he was aided by others). And Mats Strom's classic accounts (MC, p. 93) resonate with the experience of happiness and excitement in firefights, while Liander observes the need to be able to quickly transition between roles under fire.

But the new missions also intensify a development that has been going on for a long time at the tactical level: mainly, the idea that military leaders learn to "let go" of images of total control and become used to organizational life that is more ambivalent and pluralistic than in the past and within which leaders find themselves at a distance from their subordinates (Fenema, Soeters & Beers, 82010). This distance, often mediated by electronic communication, is related to what may be called distal leadership: in the midst of a mission, Ulf Henricsson (MC,

p. 75) realizes that he has to hand control down to the levels below him as he understands that they are the decision-makers on the ground:

> Now you're into something that is difficult to explain here in Sweden. Down there. I won't get anywhere if I'm not prepared to do it. I was prepared to fire and we would have done it if we had to. ... Here in Sweden it's been difficult to get people to understand how I made the assessment right when it happened.

In the new missions, this letting go is even more difficult, given that the actions of subordinates may bear strategic import. As some of the accounts in the two volumes underscore, a wrong decision by a commander may have very wide political implications.

But there is more here, as militaries are *the* model of bureaucratic organizations marked by a clear division of labor, hierarchy, standardized operations and reliance on precise regulations for achieving regularity, reliability and efficiency (Shamir & Ben-Ari, 2000). Historically, leadership in such organizations depends on formal patterns of authority and the exercise of legitimate power. Yet, as is clear from the missions that Sweden (as other countries) has been involved in during the past few decades, contemporary militaries become much more characterized by a flexible division of labor, decentralized decision-making, low reliance on formal hierarchy and greater use of informal communication between the ranks and across to other entities. The problem, as Soeters (2008) points out, is that such organizations are characterized by high role ambiguity (see mentoring relations), internal tensions (for example, among civilians and military officers) and shifting relationships (take the alliances with police forces in former Yugoslavia).

All of this means that the new missions increase the significance of leaders as the centers of gravity of combat formations. In such frameworks, leaders depend *both* on formal rank and authority (especially within their units) and on other personal bases of power. To put this point by way of example, a mentor to an Afghan battalion or an officer in charge of a civil–military mediation effort can depend no longer on position power when in contact with other entities, but rather on often slow and cumbersome consensus-building and persuasion. Indeed, it is for this reason that so many interviewees talk about the need to build, cultivate and depend on relationships. To use a phrase coined by Ogawa and Bossert (1995: 43), the "interact, not the act, becomes the basic building block of organizational leadership.

A number of points follow this understanding. First, as is evident from the volumes, even more than in the past, the leaders of combat formations provide the mental frames – the larger picture – to all of the components. The head of an ad hoc high-intensity policing force, the leader of a task force in the Congo or the person in charge of a varied staff constantly provide the meanings attributed to events by other people in ways that give their actions purpose, motivate, reinforce collective ties and direct collective action (Shamir, 1997). But this is precisely the problem as, in many of the combat formations, the blurry boundaries and the dependence of leaders on others may create problems for awarding meaning and giving purpose. It is here that the classic issues of combat units take on crucial importance, especially at the tactical level. As it comes out in the accounts, the ability of leaders to depend and motivate *their own* troops is itself a tool to move whole combat formations. Thus, in many cases, it seems that the ability of leaders to depend on the "core" element (basically, a Swedish component or unit) within the combat formation allows them to devote more attention to the other components and to impel them to action.

Second, the special characteristics of the combat formations raise questions about what may be called the temporary "psychological contracts" among participants – those informal agreements about the conditions of service and mutual obligations and duties – within them. The accounts in the two books do not reveal too much here, but there are tantalizing suggestions and intimations. To begin, those outside the military – and, at times, seconded units – may be akin to temporary staff in business firms. These elements may have much more transactional contracts emphasizing instrumental benefits at the same time as they have a much more relational contract with the "home" units to which they belong. One example is provided by Liander, who reports that the Afghan battalion commander he worked with saw him and the Swedish mentors as providing support and awarding him a high status (among his Afghan peers). In other words, the relationship was transactional in the sense of the Swedes providing resources and status symbols. Yet, in Rahmstrom's case, the fact that, after being wounded, the Afghan commander he worked with traveled from hospital to say goodbye to him indicates a much more relational contract. In any case, the task for current-day leaders is to identify these "contracts" in order to assure the performance of the combat formation.

Third, what is crucial in all of these formations is the building of relationships and, more specifically, the building of trust. This point comes out most strongly in regard to sharing: simple things such as the exchange of gossip or background information or partaking of food and drink (especially in non-formal spaces). But above all it is sharing burdens and risk. As Liander remarks, "The mission is basically of trust, and good relations are a prerequisite for this". And Hans Ilis-Alm, a special forces commander then in the Congo, observes that trust is the basis of building relationships. Yet what seems to happen is the advent of "swift trust", a form of trust that is most important in ad hoc teams, when he says, "the process gets a little quicker when you are under pressure". Accordingly, leaders of combat formations seem to master the dynamics of swift trust (Ben-Shalom et al., 2005; Hyllengren et al., 2011).

Conclusion

In this chapter, I have suggested that a fruitful way to understand current-day military leadership is to shift focus away from combat units to what I have called combat formations: those loosely structured, provisional, modular organizational forms through which most missions are now carried out. As explained, the problem in such formations is that leaders face the challenge of effecting collective and coordinated social action among varied constituent elements, each with its own goals, interests and modes of action. However, I have also emphasized that this shift does not consist of a dismissal of the older forms of military leadership. Rather, the model of change in the armed forces of the industrial democracies – and the Swedish military among them – that is suggested from the material is not a simple linear development but rather a cumulative one in which new missions, roles and environments are *combined* with prior ones. This model includes the simultaneous existence of older, more conventional roles and behaviors together with (and *not* necessarily being replaced by) newer functions and practices. Along these lines, the classic textbook units (platoons, companies or battalions) continue to be significant but also form tools or kernels for affecting, influencing the larger combat formations. To end, I add a word of caution. Although some of the skills required of military commanders are similar to those of managers of multinational corporations – negotiation, liaison, persuasion and teamwork – the very fact that the armed forces specialize in state-sanctioned use of organized violence implies that the classic traits for leading in dangerous situations continue to be of utmost importance.

Note

1 Reprinted and updated with permission: Ben-Ari, E. (2017). From leading combat units to leading combat formations: Modularity, loose systems and temporariness. In L. V. Tillberg (Ed.), *Uppdrag militar* (pp. 53–70). Stockholm: CSMS.

References

Ben-Shalom, U., Leherer, Z., & Ben-Ari, E. (2005). Cohesion during military operations? A field study on combat units in the Al-Aqsa intifada. *Armed Forces and Society*, *32*(1), 63–79.

Biehl, H. (2008). How much common ground is required in military cohesion? Social cohesion and mission motivation in the multinational context. In N. Leonhard, G. Aubry, M.C. Santero & B. Janowoski (Eds.), *Military cooperation in multi-national missions: The case of EUFOR in Bosnia and Herzegovina* (pp. 191–220). Leipzig: Socialawissenshcaftliches Institut fur der Bundeswehr.

Bury, P. (2019). *Mission improbable: The transformation of the British Army reserves.* Howgate.

Cronin, P.M. (2008). Irregular warfare: New challenges for civil-military relations. *Strategic Forum*, *234*(October), 1–12.

Davis, D.D. (1995). Form, function and strategy in boundaryless organizations. In A. Howard (Ed.), *The changing nature of work* (pp. 112–138). New York: Jossey-Bass.

Fenema, P.C. van, Soeters, J., & Beers, R. (2010). Introducing military organizations. In J. Soeters, P.C. van Fenema & R. Beeres (Eds.), *Managing military organizations* (pp. 1–14). London: Routledge.

Hasselbladh, H., & Yden, K. (2020). Why military organizations are cautious about learning. *Armed Forces and Society*, *46*(3), 475–494.

Hoffman, F.G. (2007). *Conflict in the 21st century: The rise of hybrid wars.* Washington, DC: Potomac Institute for Policy Studies.

Hyllengren, P., Larsson, G., Fors, M., Sjoberg, M., Eid, J., & Olson, O.K. (2011). Swift trust in leaders in temporary military groups. *Team Performance Management, 17*(7/8), 354–368.

Kaldor, M. (2001). *New and old wars: Organized violence in a globalized era.* London: Polity.

King, A. (2019). *Command: The twenty-first century general.* Cambridge: Cambridge University Press.

Münkler, H. (2005). *The new wars.* London: Polity.

Ogawa, R.T., & Bossert, S. (1995). Leadership as an organizational quality. *Education Administration Quarterly, 31*(2), 38–58.

Schilling, S. (2019). *Cohesion in modern military formations: A qualitative analysis of group formation in junior specialized and ad-hoc teams in the Royal Marines.* Doctoral dissertation, School of Security Studies, King's College London.

Schon, D. (1983). *The reflective practitioner: How professionals think in action.* New York: Basic Books.

Shamir, B. (1997). *Leadership in boundaryless organizations: Disposable or indispensable?* [Paper presentation]. *WORC Workshop on Transformational/Charismatic Leadership.* Tilburg: Tilburg University, 3–5 May 1997, and at the 13th EGOS Colloquium, Budapest.

Shamir, B., & Ben-Ari, E. (1999). Leadership in an open army? Civilian connections, interorganizational frameworks and changes in military leadership. In J.G. Hunt, G. Dodge & L. Wong (Eds.), *Out-of-the-box leadership: Transforming the twenty-first-century army and other top-performing organizations* (pp. 15–40). New York: JAI Press.

Shamir, B., & Ben-Ari, E. (2000). Challenges of military leadership in changing armies. *Journal of Political and Military Sociology, 28*(1), 43–59.

Shamir, B., & Ben-Ari, E. (2009). Hybrid wars, complex environments and transformed forces: Leadership in contemporary armed forces. In G.A.J. In van Dyk (Ed.), *Strategic challenges for the African armed forces for the next decade* (pp. 1–16). Pretoria: Sun Press.

Shields, P.M., & Travis, D.S. (2017). Achieving organizational flexibility through ambidexterity. *Parameters, 47*(7), 65–76.

Soeters, J.L. (2008). Ambidextrous military: Coping with contradictions of new security policies. In M. de Boer & J. de Wilde (Eds.), *The viability of human security* (pp. 109–124). Amsterdam: Amsterdam University Press.

Tillberg, P., & Tillberg, L.V. (2013). *Mission commander: Swedish experiences of command in the expeditionary era.* Stockholm: Swedish Centre for Studies of Armed Forces and Society. [MC]

Tillberg, P., Tillberg, L.V., Svartheden, J., Hamstrom, B., & Hildebrant, J. (2017). *Mission Afghanistan: Swedish military experiences from a 21st-century war.* Stockholm: Swedish Centre for Studies of Armed Forces and Society. [MA]

de Ward, E., & Kramer, E.H. (2010). Expeditionary operations and modular organization design. In J. Soeters, P.C. van Fenema & R. Beeres (Eds.), *Managing military organizations* (pp. 71–83). London: Routledge.

Winslow, D. (2002). Strange bedfellows: NGOs and the military in humanitarian crises. *International Journal of Peace Studies, 7*(2), 35–55.

Part III

Methodologies for the Study of Military Formations

6 Research approaches to the study of combat formations

A Personal Note

Uzi Ben-Shalom

Introduction

There is a basic assumption that the military is an institution notorious for being hierarchical, bureaucratic and replete with prescriptive procedures. In contrast to this often justified view of the military as in institution, military operations themselves are often perceived by those involved as fluid and even chaotic. A primary reason for this contradiction is the need for the adaptation of the military structure to operations in an environment characterized by great uncertainty (Soeters & van Fenema, 2010). Military missions by their nature are subject to the uncertainties of combat. As part of the process of operation, there is a high possibility for the amalgamation of units into combat formations. The reasons for this mode of action are often planned in advance according to doctrine, as in the case of a combined-arms brigade. But, in many cases, they result from emergencies and surprise or ad hoc practical necessities. A good example is the ad hoc linkup of ground and air components while coordinating the rescue of wounded soldiers (Ben-Shalom & Tsur, 2018). Another example is the formation of a new unit composed of the survivors of units that have suffered heavy losses (Rush, 1999). Under these conditions, the military is forced to combine various elements and work in a "mission formation type of action". My contention is that mission formations are emblematic of such fluidity and that this aspect of the military institution requires an appropriate approach by behavioral scientists.

Historically speaking, the great wars in the 19th century were already fought by state coalitions. However, mission formations have gradually become a prerequisite for any military operation and a vital requirement in combined-arms ground combat (King, 2010). The military sociological study of this process seems to lag behind this gradual development. The difficulties of doing a study during an operation are reflected in what is being studied. There is far less information about force employment (as during skirmishes or the execution of offensive or defensive operations) compared with force generation (as in facilities of recruitment and training or among veterans, etc.). In a recent literature review, Soeters (2018) tracked the roots of general sociology in the study of the military. It seems that most sociologists do not approach the subject of military action (which is very often conducted by a mission formation), but rather focus on the military institution itself. In this way, the study of military action is rather neglected in the field of sociology.

Challenges and requirements

It is my contention that conducting research on mission formations requires both access to military units themselves and a professional knowledge of the military. If the research team does not possess such access or knowledge, it will have difficulties in creating an appropriate research design. If that access or knowledge is limited, the team will likely fail to collect the relevant information, and, even if that information is collected, it will likely be misunderstood. My aim is to present some of these necessary prerequisites for behavioral scientists who set themselves the goal of studying military action, in which combat formation is likely to occur.

The study of mission formation requires a multidisciplinary team as the formation often has roots in various sub-elements. It is a question of access and design. Take, for example, the case of today's maneuvering Israel Defense Forces (IDF) Ground Forces battalion. This unit operates a combination of reservists, regulars and conscripts who are all taking part in a single maneuvering ground battalion. This unit commands armored vehicles, combat engineering equipment, surveillance assets and communication equipment. It may be supported by artillery, air and electronic detachments. In order to operate seamlessly, this battalion requires the integration of skilled technicians and, at times, even civilian technical experts. If it is to maneuver inside urban terrain, it will need to have additional components such as spokespersons and humanitarian and, sometimes, legal experts. The complexities of this amalgamation require knowledge on the part of the researcher and their team concerning each of these components and a research design that allows for the collection of the necessary information.

These challenges could be resolved by training and educating today's professional military sociologists. As military sociologists, we need to make ourselves familiar with the building blocks of these formation measures by reading military doctrine, visiting the training of these formational units and discussing with personnel who are involved in creating and supporting them. Experiments with the future design of formations are also important. But such an educational process does not negate the "usual" challenges of military research. Such obstacles include, among others, the problematic alliance between the researcher and the military institution, the commander of the unit and its personnel. Such obstacles are often presented in personal accounts of military scholars (Gazit & Maoz-Shai, 2010; Ben-Ari & Levy, 2014). Studying the actions of such formations can be demanding, as the research team must not only have access to the field but also have the ability to support itself while deployed. How can we demand such a capability from military sociologists?

A personal note

After my compulsory service in the IDF Intelligence Branch, I went into the reserve component. This is a very common process in Israel, where growing up means years of compulsory military service (which in my case lasted 5 years). But then, 10 years after my initial draft, while studying for an MA in the social sciences, I re-enlisted in the IDF as an expert in social research. I subsequently served for 20 more years in a research position within the IDF personnel branch and later in the Ground Forces Command, where I still am an active reservist. While in uniform, I always kept close ties with academia, completing my MA, PhD and Post-Doc while still a serving military officer. I was always privileged in getting access to military actions, though, while in uniform, I made the theoretical analysis secondary to the practical needs of the military undertakings for which I was responsible.

At first, I was a human resources researcher but gradually became more interested in military action. This began as a personal interest and not something I planned. During my service, Israel had a series of military undertakings: the possibilities of rocket attacks by Iraq; the guerilla warfare against Hezbollah in Lebanon and the subsequent withdrawal from the Lebanese Security Zone; combat in Gaza and the disengagement from Israeli settlements there; the Second Intifada; the Second Lebanon War; and now the Third Intifada. I was involved in research in each and every one of them. The Ground Forces had to adapt themselves each time, and my research was involved in these processes (Ben-Shalom, Lehrer & Ben-Ari, 2005; Ben-Shalom & Fox, 2009). Among the lessons I learned during my varied service was that one should adapt oneself to use any useful research method and avoid clinging to any theoretical commitments. As I have been studying military undertakings since the mid-90s, I have also concluded that the behavior of humans in combat should never be taken for granted and requires, above all, a genuine interest by the researcher. I believe this notion would be worthwhile for anyone conducting research on mission formation in combat.

Research approach

Military activities are innately challenging owing to their chaotic nature. While in combat, participants experience a diversity of viewpoints and intense emotions. For example, during one research study, I had the opportunity to interview a combat platoon following a painfully unsuccessful skirmish. One of the things that emerged was that each and every one of them had a different view of the attack and the way it developed. Intense emotions were involved, and arguments over the effectiveness of the skirmish were not resolved immediately after the fight, or later. The complexities of combat situations are subject to personal interpretation, and this can bewilder the research team. At times, it is a gradual and long-term process of sense making (Ben-Shalom & Shamir, 2011; Padan & Ben-Shalom, 2019). The complexity of mission formation can enhance such difficulties, and the military's own formal debriefing process – which is heavily centered on questions of tactics and communication – is not all that interested in such personal contradictions. Moreover, mission formations are usually created in theaters of operation and during active combat situations or deployments, and, as a result, the restriction of access is a common obstacle. Ironically, such access is often granted after a problem is identified in the military function. Indeed, militaries have now begun to allow access to social scientists as part of a remedy and as part of the process of formal investigation (for example, see Winslow, 1999; Elsey, Mair, Smith & Watson, 2016). This suggests that most of the small yet growing literature on mission formations is based on these formal debriefings.

The military sociological approach to military action is often grounded in a set of research tools which focus on both qualitative and quantitative approaches (Carreiras & Castro, 2012; Soeters & Shields, 2014). Williams, Jenkings, Woodward and Rech (2016), for example, clustered scholarship in military research into four categories of "text", "interactions", "experiences" and "senses". This intriguing combination of perspectives is informative, but was not focused on action and certainly not on mission formation. In my view, such a distinction, although very intuitive and descriptive, may lead to a kind of artificial cleavage between field observations and post-event research. The ability to study military actions using just one research approach is self-defeating. Simply put, the research team is likely to be surprised by the conditions in the field and will have problems in articulating a research question. The lack of a clear research question and research tools to answer that question presents serious obstacles to the study of military action. On the more quantitative side, of course, the problem becomes even more apparent. Here, the fluidity of the operational environment defeats the scientific techniques of sampling, comparison and well-validated measurement tools. One of the remedies for these challenges is the diversity of the research team. Such a working team is essential to surviving in operational areas as it ensures a greater ability to cover more areas of research activity along with the appropriate tools of analysis.

The accumulating research experience of the decades since the first Gulf War has contributed to the body of research on the military. Here, the research method of choice in terms of mission formation is less of an issue than the planning itself. It is the state of mind that is more relevant, and as a rule of thumb the research team – and a team is a prerequisite – must be capable of collecting as much primary and secondary information as possible. In the words of Ruffa and Soeters:

> comparing military actions by national contingents in an appropriate manner requires a careful research design, in which some of the research variables are identical and others are not. Besides, cross-national research needs highly qualified researchers who know how to apply similar instruments of data collection in different languages to equivalent samples in a manner that excludes preconceived ideas and ethnocentrism as much as possible.
>
> (Ruffa & Soeters, 2014, p. 225)

In my mind, what makes all the difference is the curiosity of the researcher about behavior in combat, coupled with the prerequisite knowledge of military institutions and activities. But a combination of academic and practical knowledge is not the only prerequisite. As mission formation is studied, it is likely to include the violence, killing and death of close-quarters combat. And a study of these issues often involves a completely different approach to military actions. While in garrison, the observer is likely to see order, yet, when facing death, such order may be achieved by different measures. Therefore, improvisation on the part of the researcher is an essential part of the study of mission formation, and this can only be fully achieved at the conclusion of any given operation (Ben-Shalom, Klar & Benbenisty, 2012). These issues imply that any research on mission formation cannot be based on the general theory of behavioral sciences that is disconnected from the praxis of the military institution itself. This requires specialization in military organizations, from familiarity with the gatekeepers of military units, the leadership, to an understanding of the jargon used (Muller-Wille, 2014). Typically, this is to the advantage of those academics who were drafted into – or otherwise served in – a military organization and who were involved in military actions (Jans, 2014). Accessibility to military environments enables a deeper understanding of the complexity of the military organization. As a member of the seminal "American Soldier" project contemplated:

> one could not stay with a front-line combat organization over periods of months without receiving indelible impressions of the impermanence of specific persons and the crucial importance of the formal organization of authority, supply, support, and coordination.
>
> (Williams, 1984, p. 186)

This acknowledgment recognizes the immense complexity of military action and the role of formal control by the command structure. Accessibility to the social dimensions of operations also allows a more realistic and critical view of the military. Other hallmarks of military sociologists who studied the military organization were important simply owing to the embedded nature of their researcher. Little (1964) reported on the racial and national composition of a rifle company in Korea, and Moskos (1975) identified the differences between short-timers and new recruits in Vietnam. Much more recent analysis is centered on the combination of mixed-gender infantry units (Ben-Shalom, Lewin & Engel, 2019). At the same time, these analyses were generally directed toward the social structure of a single branch combat team – most often an infantry unit – whereas the challenges of mission formation are somewhat different, more complicated and call for a different approach.

Key questions

In order to study mission formation, researchers may be reflexive about following guideline questions while conducting their work: "What?", "Who?" and "How?". All three are actually embedded in any research of action in a diversity of settings, but what makes them unique here is the military organization itself and its goal in inflicting and sustaining violent action. The reader should note that the bulk of military sociology knowledge on the following guidelines is centered on other aspects of military action.

What is the mission?

A mission is the emblem of military action. It is imperative and anchored in doctrine. Yet, once they are set in motion, missions are most often subject to a diversity of interpretations, especially when the enemy executes their own plan amid the "Fog of War". Therefore, the researcher should try – to the best of his/her ability – to separate planned missions from those which result from enemy activity.

Who is involved?

A military mission is conducted by many interdependent branches. Armored brigades will not advance without military engineers, heavy vehicles and knowledge of the terrain. An air assault will not occur unless some air transportation will link with the ground element and it must be coordinated with artillery and supply. Therefore, a study of such conventional mission formations must include awareness of who the agents are that are being linked to address the mission. Who are the soldiers, leaders, units, branches and services involved in the formation? In the near future, as robots may be used in the battlefield, military sociologists may expand their interests into human–robot interaction in mission formations.

How is it done?

Mission formation is not trivial or taken for granted – it is but the end of a force generation process. The research team needs to be aware of varied processes that link the various parts together: their doctrine, operational concept, command and control, training and equipment. The communication assets are crucial in deciphering the process through which mission formations are both created and function.

A research design

The research – which should be conducted by a team with solid military knowledge – is very often conducted post hoc, as the ability and interest in ad hoc formation becomes clear only after a battle is concluded. Of course, the interest commanders may have in the research of specific battles becomes apparent only in time. While a mission is under way, the interest and ability to allow formal research is scant, and the willingness to allow it is not taken for granted. When the basic question is understood – and often it is a process and not a clear-cut definition – the research approach could be decided.

This is an ongoing research process and is usually marked by an accumulation of knowledge during or after the action. It also involves greater mobility of the research team, who often find themselves collecting information from various sources and then linking together those sources to analyze it and set a new goal for ongoing research. It is wise to be very flexible with the analytical tools and time structure. At times, a hasty response is crucial, and a representative of the research team should immediately link to a formation assembly area or commander debrief.

In other cases, a deep understanding of what happened may become clear only long after the conclusion of a battle. While in the field, good connections with quartermasters and other commanders can make all the difference in securing appropriate places to stay, work and even eat. Let me briefly explain this contention by using my own research experiences in different levels of action.

Case studies

Micro-level examples

Military formation is the action of different parts that are linked in order to achieve a certain mission. At the micro-, or techno-tactical, level, this roughly equates to units below the level of a company – the performers are subject to the conditions of close-quarter violence. Their action is not entirely coordinated by military staff or by the formality of doctrine but by the immediacy of close combat leadership. Time and again I have found that the behaviors of people in combat may differ substantially from the expectations of doctrine.

The first example is highly relevant to certain skirmishes in Israeli urban areas in the last decade. The combat formation is completely arbitrary and is composed ad hoc by an amalgamation of soldiers, policemen and security personnel facing terror attacks inside Israel. Here is a typical example based on a previous study (Ben-Shalom, Moshe, Mash & Dvir, 2019):

> On October 9, 2015 in the Israeli city of Afula, an Arab-Israeli woman waved a knife towards civilians and uniformed security persons and policemen, who eventually shot and seriously wounded her. This security event occurred in the midst of a terror wave in Israel, which is now labeled, "The Intifada of the Individuals". The event was filmed by a pedestrian and quickly went viral, provoking much public debate. The video revealed extreme peculiarities of the woman attacker and the security responders. At first, the event was considered a terror attack, but was later labeled as a copicide; that is, suicide by provocation of the police. The behavior of one specific female soldier in this event also attracted wide attention by commentators. Filmed from behind her shoulder, she is seen holding her rifle aimed at the threatening Arab knife-woman. Even though the camera movement was shaky, it could be clearly discerned that the female soldier was apparently holding a lollypop in her hand. During the incident, the media focused on the fact that the lollypop fell to the ground and as the soldier advanced, she picked it up, continuing to aim her rifle at the attacker. However, she and other female soldiers refrained from shooting the knifewoman, although a crowd was heard shouting "Kill her!" and "Shoot her!".

Such an occurrence entails a certain low-level mission formation that comes together to confront the attack. It very often includes civilians and pedestrians who rush to the event, especially after the attack is concluded.

What was the mission? There is the need to stop a terror activity.

Who is involved? Directly involved are policemen, conscripts, border patrol female combatants, civilian security and pedestrians.

How was it done? Imitation of nearby policemen and self-selection.

A research design: Monitoring video material and in-depth interviews.

Monitoring dozens of videos of this kind of activity, I have found that the ability to form an effective combat formation rests – again – on close combat leadership at the site of the incident. The behaviors and ability to work together mean that the researcher must follow sounds, gestures and eye contact made by the members of the temporary team.

Here is another common combat formation which is a typical example of a skirmish along Israeli borders in 2012:[1]

> The terrorist squad included three combatants heavily armed. They probably spotted the IDF force beforehand and left some of their arms in a small cave while climbing the hill for observation. The IDF trackers arrived just as they were preparing themselves. They must have been interrupted seeing a full platoon of an IDF artillery battalion arriving as replacement. The newly arrived soldiers made a lot of noise as they were busy unloading the gear and prepared themselves to replace the platoon of light-infantry soldiers from a mixed-gender battalion. They had by now been watching for over three days a group of refugees from Africa who illegally crossed the Israeli border with Egypt. These refugees were crouching under a shade in a small unnamed dry brook in the afternoon sun. The onslaught of firing tore the silence of the mundane desert. The ensuing minutes are a mixture of voices, shooting, killing and heavy explosions. The formation of soldiers that fought the attack never meant to operate the way it did. It was forced out of necessity and was made up of self-selected leaders and followers. It eventually ended in three soldiers from both units – composed of both genders – who stormed the enemy position. During the skirmish they killed the terrorist squad while losing one soldier.

Three years later, I tried to make sense of the recollections of the soldiers and commanders. I devised a combination of research approaches including in-depth interviews, drawing of maps and making comparisons with the official debriefings and, finally, presented the results to senior commanders. Having in mind vague information – most of it incorrect – from the open media, I drew the following composition of participants in a way that made sure that I would cover every part of the structure. The allocation of the interviews required a snowball sample and patience, as soldiers were by then out of compulsory service. In all, I needed ten interviews to cover this event. I then made a comparison with three similar incidents, and, by then, I was sure that I had in mind a pattern of behaviors, decisions and social processes during and after the fight (Figure 6.1).

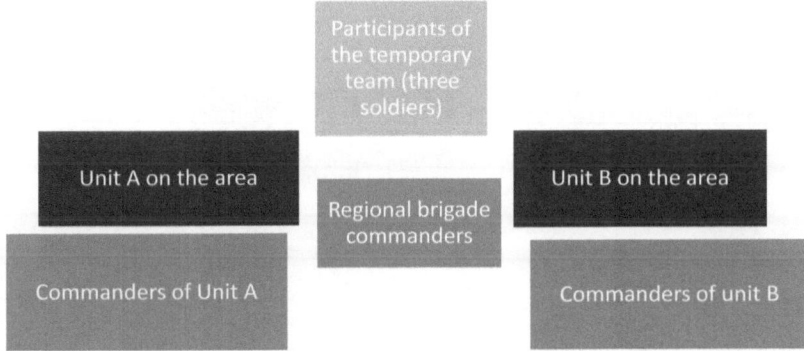

Figure 6.1 A sketch of a mission formation during a squad-level skirmish. *What was the mission?* Protecting a border turned into destroying an enemy position. *Who was involved?* Directly involved were members of two different units. *How was it done?* Formal procedures of replacement turned into self-selection process. *Research design*: In-depth interviews with all personnel involved from both units and the regional brigade. Drawing a sketch of battle space from recollections of participants and review of combat debriefings as a base line for the debriefing process.

One of the important lessons I drew was the fact that combat execution requires a series of self-selection processes in which not everybody involved is taking an active part, and such a selection process exceeds presumption of gender differences and formal authority (Ben-Shalom, Klar & Benbenisty, 2012; Ben-Shalom, Lewin & Engel, 2019).

Meso-level example

The following case study is a mission formation at the meso level. It is based on a study of air and ground forces performing ground-centric operations (Ben-Shalom & Tzur, 2018). The research team included experts from both ground and air forces who had specific technical experience in both services. The research was conducted using in-depth interviews that covered each side's general perceptions of the actions and the cooperative ability of the other. The following is an example of how ground forces personnel perceive the activities of the Air Force:

> There is almost no situation in which dead or wounded are involved and the risk is higher than average, in which the pilots don't attempt to carry out their mission in the best possible manner, and not only in recent years. I can't remember any incidents complicated by dead and wounded in which a request was submitted to the AF that was not put into practice – the priorities here are very clear and they will take any risk to save the lives of IDF soldiers. (In training) it appears as if they are not prepared to take unnecessary risks, they are extremely prudent and cautious in their maneuvers with us – much too cautious.
>
> (Brigade XO; Figure 6.2)

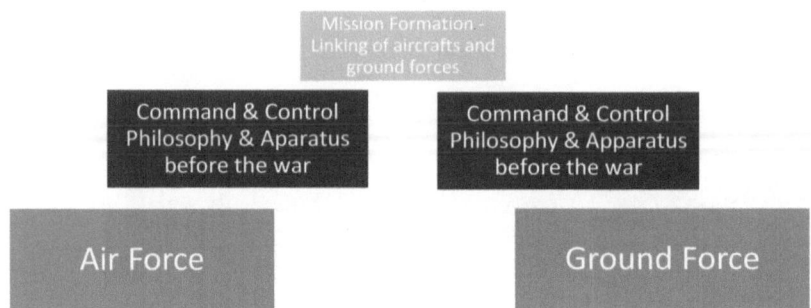

Figure 6.2 A sketch of an air–ground mission formation. *What was the mission?* Air and ground forces joint operations. *Who was involved?* Medium-level commanders (majors to colonel). *How was it done?* Formal procedures versus informal cultural scripts. *Research design*: In-depth interviews with commanders from both services.

This research was based on about 40 in-depth interviews. The ability to analyze and publish materials for such an activity is limited, and therefore in-depth interviews were selected. The research team identified key themes in the narratives of commanders of both services. One of the lessons that the research team recognized was that of the formal doctrine and service procedures actively being interpreted during the action itself. Perception, then, was influenced not only by regulations and experience but also by personal acquaintances and unit reputations.

Macro-level example

The disengagement from the Gaza Strip and the Northern West Bank in the summer of 2005 was an internal security operation conducted by a coalition of military and the Israeli police forces. The research team I was responsible for covering the force generation and deployment and the dismantling of brigades that were responsible for the evacuation itself. This research was conducted for the purposes of assessing the well-being of the soldiers who executed the often painful evacuations (Ben-Shalom, Knafo & Goldner, 2014).

The research was conducted in the 2 months preceding the Israeli withdrawal from the Gaza strip, and during the following 8 months, among healthy soldiers, hand-picked from their original units to participate in the Gaza withdrawal task. In spite of the fact that the Gaza withdrawal was not a combat mission, but rather a peacekeeping or policing mission, which is less likely to cause psychiatric problems, the many stressors involved in this mission were expected to lead to differential adjustment among the soldiers. The research team was gathered to collect information in the military and police organizations, which took part in all strategic aspects of the operation. This included more than 60,000 soldiers. The research was based on the collection of impressions on the ground through questionnaires which were submitted to all participating units (Figure 6.3).

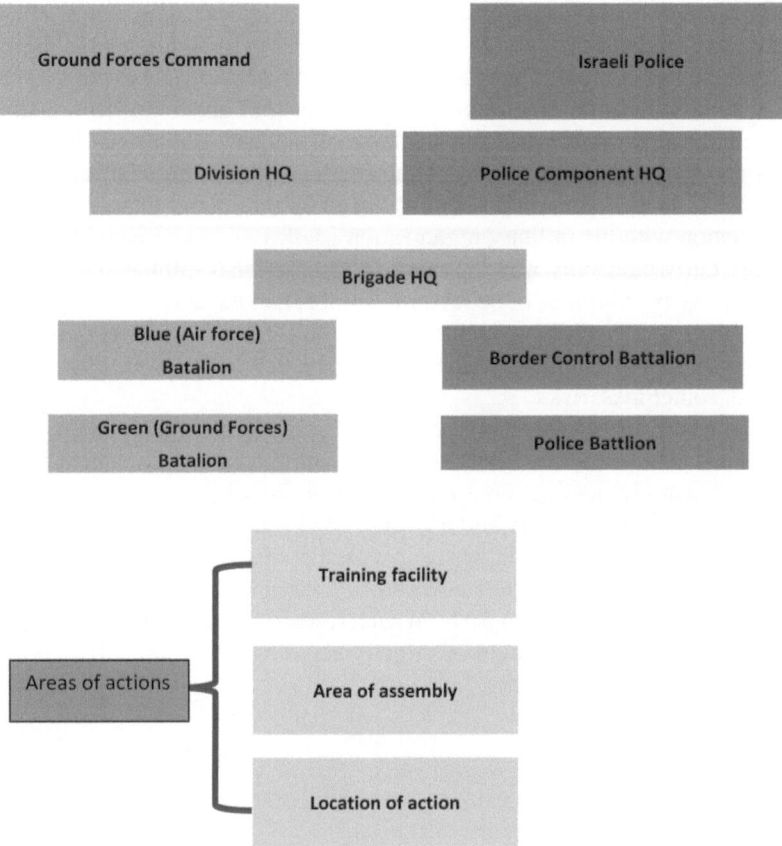

Figure 6.3 A sketch of a mission formation during the disengagement from Gaza in 2005, organized top–down. *What was the mission?* Internal security operation by national-level military and police formations. *Who was involved?* Soldiers and leaders in ad hoc formations. *How was it done?* Two months training of units and deployment at battalion level. *Research design*: The research design included dozens of military psychologists deployed with the forces as soon as they were formed, as well as representatives from the IDF and Israeli Police.

The research team followed the training and organization of the units closely. The data were collected by a myriad of sources, but most importantly by field observations which were discussed every evening (where possible). The minute processes of the forces on the ground were of great interest to senior commanders who were also present on the ground in temporary command posts.

Conclusion

This short chapter has covered my own reflections on the study of mission formation. Once decided – or ordered – to do research on a mission formation, the researchers ought to proceed accordingly. The research should be conducted by a multidisciplinary team, using a variety of methodological approaches, after a comprehensive study of military knowledge and sensitivity to the human dimension in the study of combat. The research design should allow access to and collection of the information required, without any overall adherence to pre-study research methods. In a recent account, the Israeli military sociologist Nir Gazit (2019) contemplated how to overcome the challenges of doing anthropological studies in the military. Gazit reflected on the need for reasonable expectations and for coordination with the military organization itself. I would agree with Gazit's contentions but would only add the need for professional military knowledge, awareness of the peculiarities of behavior in combat and the need for an appropriate research design when doing a study of mission formation. Last but not least, doing research on a mission formation is arduous fieldwork subject to great constraints and, sometimes, risks.

Finally, I would also argue that the study of mission formation is perhaps better understood as a combination of a state of mind (willingness to undertake the challenge of such a study) and the constant interest of the scholar (which will drive him/her to study military history and doctrine). If you are a military sociologist in uniform, there are always other things to study rather than combat. And our adherence to a fixed research method – qualitative or quantitative – may also block our genuine interest when striving to perform such research. But the study of military action is sorely needed in the field of military sociology (Soeters, 2018), within both the academy and the military.

Note

1 www.ynet.co.il/articles/0,7340,L-4324466,00.html (Hebrew).

References

Ben-Ari, E., & Levy, Y. (2014). Insider/outsider perspectives. In J. Soeters & P.M. Shields (Eds.), *Routledge handbook of research methods in military studies* (pp. 9–18). New York: Routledge.

Ben-Shalom, U., Lehrer, Z., & Ben-Ari, E. (2005). Cohesion during military operations: A Field study on combat units in the Al-Aqsa Intifada. *Armed Forces & Society, 32*, 63–79.

Ben-Shalom, U., & Fox, S. (2009). Military psychology in the Israel Defense Forces: A perspective of continuity and change. *Armed Forces and Society, 36*(1), 103–119.

Ben Shalom, U., Klar, Y., & Benbenisty, I. (2012). Characteristics of sense making during combat. In J.A. Lawrence & M.D. Mathews (Eds.), *Handbook of military psychology* (pp. 218–230). London: Oxford University Press.

Ben-Shalom, U., Knafo, A., & Goldner, I. (2014). The role of internal locus of control in coping with anticipatory and post-event stress among IDF soldiers. *Military Behavioral Health, 2*(1), 18–25.

Ben-Shalom, U., Lewin, E., & Engel, S. (2019). Organizational processes and gender integration in operational military units – An Israel Defense Forces case study. *Gender Work and Organizations.* doi:10.1111/gwao.12348

Ben-Shalom, U., Moshe, R., Mash, R., & Dvir, Z. Amit. (2020). Micro-sociology and new wars – Visual analysis of terror attacks during the "Intifada of the Individuals". *Armed Forces and Society, 46*(2), 281–301.

Ben-Shalom, U., & Shamir, E. (2011). Mission command between theory and practice – The case of the IDF. *Defense and Security Analysis, 27*(2), 101–117.

Ben-Shalom, U., & Tsur, Y. (2018). Scripts of service culture – A hidden dimension in the successful joint operations of the Israeli Air Force and Special Forces. In E. Ben-Ari, K. Michael & J.G. Turnley (Eds.), *Special operations forces in the 21st Century: Perspectives from the social sciences* (pp. 106–119). NewYork: Routledge.

Carreiras, H., & Castro, C. (Eds.). (2012). *Qualitative methods in military studies: Research experiences and challenges.* New York: Routledge.

Elsey, C., Mair, M., Smith, P.V., & Watson, P.G. (2016). Ethnomethodology, conversation analysis and the study of action-in-interaction in military settings. In A.J. Williams, N. Jenkings, R. Woodward & M.F. Rech (Eds.), *The Routledge companion to military research methods* (pp. 180–195). New York: Routledge.

Gazit, N. (2019). The entanglements of military research in home and abroad: An experience of an Israeli anthropologist. In B.R. Sorensen & E. Ben-Ari (Eds.), *Civil-military entanglements: Anthropological perspectives* (pp. 251–273). Berghahn Books.

Gazit, N., & Maoz-Shai, Y. (2010). Studying-up and studying-across: At-home research of governmental violence organizations. *Qualitative Sociology, 33*(3), 275–295.

Jans, N. (2014). Getting on the same net: How the theory-driven academic can better communicate with the pragmatic military client. In J. Soeters, P.M. Shields & S. Rietjens (Eds.), *Routledge handbook of research methods in military studies* (pp. 39–48). New York: Routledge.

King, A. (2010). The internationalization of the armed forces. In J. Soeters, P.C. van Fenema & R. Beeres (Eds.), *Managing military organizations: Theory and practice* (pp. 60–72). New York: Routledge.

Little, R.W. (1964). Buddy relations and combat performance. In M. Janowitz (Ed.), *The new military – Changing patterns of organization* (pp. 195–224). Sage.

Moskos, C.C. (1975). The American combat soldier in Vietnam. *Journal of Social Issues, 31*(4), 25–37.

Müller-Wille, B. (2014). Doing military research in conflict environments. In J. Soeters, P.M. Shields & S. Rietjens (Eds.), *Routledge handbook of research methods in military studies* (pp. 60–72). New York: Routledge.

Padan, C., & Ben-Shalom, U. (2019). Sensemaking of military leaders in combat and its aftermath – A phenomenological inquiry. *Political & Military Sociology – An Annual Review, 46*(2), 324–342.

Ruffa, C., & Soeters, J. (2014). Comparing operational styles. In J. Soeters, P.M. Shields & S. Rietjens (Eds.), *Routledge handbook of research methods in military studies* (pp. 216–228). New York: Routledge.

Rush, R.S. (1999). A different perspective: Cohesion, morale, and operational effectiveness in the German army, Fali 1944. *Armed Forces and Society, 25*(3), 477–508.

Soeters, J. (2018). *Sociology and military studies: Classical and current foundations.* New York: Routledge.

Soeters, J., Shields, P.M. & Rietjens, S. (Eds.). (2014). *Routledge handbook of research methods in military studies.* New York: Routledge.

Soeters, J. & van Fenema, P.C. (2010). Introducing military organizations. In J. Soeters, P.C. van Fenema & R. Beeres (Eds.), *Managing military organizations: Theory and practice* (pp. 19–32). New York: Routledge.

Williams, J., Jenkings, N., Woodward, R., & Rech, M.F. (Eds.). (2016). *The Routledge companion to military research methods*. New York: Routledge.

Williams, R.M. (1984). Field observations and surveys in combat zones. *Social Psychology Quarterly, 47*(2), 186–192.

Winslow, D. (1999). Rites of passage and group bonding in the Canadian Airborne. *Armed Forces and Society, 25*(3), 429–457.

Part IV
Glocalized Mission Formations

7 Institutional isomorphic change in South Korea's UNPKO mission formation

Insoo Kim and Young-Il Choi

Introduction

After the end of World War II, "states have ceased fighting each other over disputed territory ... and armies concentrate increasingly on the repression of civilian populations, the combat of insurgents, and seizure of power" (Tilly, 1992, p. 203). Along with a decrease in interstate war, the nature of war rapidly changed to the so-called New Wars or Hybrid Wars, which refer to "a mixture of war, crime, and human rights violations" (Kaldor, 1999, p. 11). In the New War, "the strong may be weak, and the weak, paradoxically strong" (Vlahos, 2003, p. 7). Because there is no such thing as the enemy center of gravity suggested by Carl von Clausewitz (2008, p. 485) in New Wars, the way of fighting was different from the past. Consequently, the armed forces of industrial democracies, trained and educated to fight the interstate conflict, necessitated a new way of mission formation that it had never experienced before.

The South Korean Army was no exception. It grew up with inter-Korean conflicts. The fundamental military mission to deter, fight and win a conventional war in the Korean Peninsula never changed, even after the end of the Korean War (1950–1993), because North Korea conducted more than 3,000 military provocations against the South (Republic of Korea Ministry of National Defense, 2016, p. 288). In the early 1990s, when the United Nations (UN) first asked to take part in the international peace effort, South Korea was unfamiliar with peacekeeping activities in civil war countries. In the process of repeating the United Nations Peacekeeping Operations (UNPKO), however, the organizational structure of the UNPKO units has converged to the same pattern of integrating all functional elements into one unit.

This study asks how South Korea's UNPKO mission formation has become as it is today. By the mission formation, we mean action sets in which several functional elements work in conjunction with one another with some degree of regularity. The concept of the mission formation, suggested here, is substantially used to examine the homogeneity of how to organize the UNPKO units. Two types of homogenization can be attributed to this change: competitive isomorphism and institutional isomorphism. First, organizations in the same sets of environmental conditions resemble one another because those organizations inefficiently acting in a given situation will be ruled out (Hannan and Freeman, 1977, p. 939). However, mission formation does not necessarily reflect a rational choice for greater efficiency because organizations tend to seek legitimacy (DiMaggio & Powell, 1991, p. 65).

This chapter presents a neo-institutional explanation of the dynamic behind the changes in South Korea's UNPKO mission formation. The following section reviews the historical development of South Korea's overseas peace operations. Then, we discuss why competitive isomorphism is not suitable for analyzing the organizational convergence of the UNPKO units. The next section demonstrates that the UNPKO mission formations began to look like each other "as they respond to similar regulatory and normative pressure, or as they copy structures adopted by successful organizations under conditions of uncertainty" (Orrù et al., 1991, p. 147). Finally, we conclude by considering how our findings contribute to discussions about the dynamic of mission formations.

Change in organizational structure of UNPKO units

A brief history of the South Korean Army

After the end of World War II in 1945, the US Army Military Government occupied the southern part of the Korean peninsula. It established one constabulary regiment called the South Korean Defense Guard to support the police. Along with the establishment of the government in 1948, South Korea renamed it the Republic of Korea Army and increased its size to 100,000 troops in eight divisions. The South Korean Army remained mostly a constabulary force until the breakout of the Korean War. The Korean War started with North Korea's surprise attack on 25 June 1950, and ended with the armistice agreement on 27 July 1953, leaving substantial human and material damage in its wake. Because South Korea's national power was inferior to North Korea's in the 1950s, South Korea could not successfully deter North Korean threats without military support from the US. To prevent the expansion of communism into East Asia, the US government armed and trained the South Korean Army. Consequently, it rapidly developed to 660,000 troops in 18 divisions through the Korean War.

In the 1960s, the conflict in Vietnam was heating up. South Korea was worried about the US moving its infantry divisions in South Korea to Vietnam. In this situation, it agreed to send combat forces to Vietnam when the US asked it to do so. Two infantry divisions and one marine division fought alongside South Vietnam until the end of the Vietnam War in 1973. Through the Vietnam War, the South Korean Army grew into a more capable and experienced war-fighting machine. As South Korea's defense industry began producing sophisticated weapon systems in the 1980s, South Korea could successfully build up a self-reliant defense posture strong enough to reduce the likelihood of engaging in a conventional war.

In the early 1990s, the collapse of the communist states paved the way for the two Koreas to change inter-Korean rivalry and thus join the UN at the same time, in 1991. Since then, the UN has continued to ask South Korea to participate in UNPKO missions, where increasing civil war and the resulting violation of human rights demanded new tasks that the South Korean Army had rarely experienced before. At the request of the UN, the South Korean Army began engaging in peacekeeping operations, humanitarian assistance and disaster relief efforts abroad. In the 2000s, its involvement in the UNPKO has expanded significantly for two reasons. First, after the attack of 11 September 2001, the US and its allies shifted their concern to the War on Terror. As the UNPKO run by the advanced western countries diminished, South Korea wanted to make use of UNPKO for a higher international profile. Second, it enacted a law relating to UNPKO in 2010 so that it could respond to UN requests quickly. This law enabled the government to discuss whether to participate in UNPKO before getting the consent of the National Assembly.

South Korea's contemporary UNPKO missions

The South Korean Army's field manual defines all overseas military missions as a peace operation and categorizes them into three groups (Republic of Korea Army, 2009, pp. 1–14). The first is UNPKO, in which South Korea has participated seven times since 1993. Starting with the 189th Engineer Battalion dispatched to Somalia in 1993, South Korea sent the Western Sahara Medical Assistance Group (MAG) to Western Sahara in 1994, the 101st Field Engineer Group to Angola in 1995, the 522nd Peacekeeping Group to East Timor in 1999, and the Haiti Reconstruction Support Group to Haiti in 2010. Currently, the Lebanon Peacekeeping Group and the South Sudan Reconstruction Support Group are working.

Second, the Multinational Forces Peace Operation (MNFPO) refers to conflict resolution, peacebuilding and reconstruction support led by a regional security organization or a specific state. South Korea has conducted this type of peace operation eight times in Kuwait, Afghanistan and Iraq. At the outbreak of the Gulf War in 1991, the US called on South Korea to take part in the UN sanctions against Iraq. South Korea dispatched the Gulf War MAG to Saudi Arabia. After the 9/11 terror attacks in 2001, it organized the 924th MAG in 2001 and the 100th Construction Engineer Group (CEG) in 2003 to support the US war efforts in Afghanistan. At the request of the US in 2003, it organized the 320th MAG and the 1100th CEG and sent them to Iraq. In 2004, it reinforced two civil affairs brigades, mostly consisting of paratroopers, and established the Zaytun Peace and Reconstruction Division to integrate all units in Iraq under unitary command. In 2008, it replaced the 924th MAG and the 100th CEG in Afghanistan with a civilian provincial reconstruction team (PRT) and dispatched an infantry unit to support the PRT activities in 2010. Recently, one navy destroyer has taken part in Operation Enduring Freedom – Horn of Africa, in Somali waters.

Finally, defense exchange and cooperation (DEC) refers to military training assistance and disaster relief assistance provided at the request of the country concerned. In 2013, South Korea sent the Joint Support Group to the Philippines to provide disaster relief. One special force unit has trained the UAE Special Force at the request of the Crown Prince of UAE since 2010.

Table 7.1 provides a comparative overview of South Korea's peace operation units. First, the unit size and per capita budget of a peace operation unit differ markedly according to the type of peace operation. MNFPO units embrace the largest size and per capita budget, with an average of 668 people and US$89,100. UNPKO units, in contrast, have only an average of 259.5 people and US$44,700. The South Korea–US alliance can best explain the difference in size and budget because South Korea has more actively participated in the MNFPO led by the US than other peace operations owing to the South Korea–US military alliance.

Organizational homogeneity of UNPKO units

Although all three types of peace operation units carry out missions abroad, we can find a clear difference in the organizational structure of UNPKO units. According to the military doctrine of South Korea, a battalion is the primary military unit capable of independent operation, so that it is homogeneous in terms of its functional elements. Although it is possible to reinforce a combat battalion temporarily with other smaller units from different army branches, there is no such concept that infantry, engineer and medical units work in conjunction with one another in a battalion. Accordingly, in the early 1990s, South Korea did not newly organize UNPKO units but selected a unit that best suited the purpose of distich. For example, the 189th Engineer Battalion and the 101st Field Engineer Group are the original names of each unit.

In late 1990, this pattern changed in a way that integrated more individuals from different army branches. Even though the proportion of each subunit varies over UNPKO units according to the overall mission, Table 7.2 shows that UNPKO units consisted of the personnel from all three functional elements. For example, the 101st Field Engineer Group consisted mostly of engineers but also had an infantry platoon and a medical subunit. The 522nd Peacekeeping Group was composed chiefly of infantries, but 179 of 419 troop members at the headquarters (42.7 percent) were personnel from combat service branches and combat service support branches. The Lebanon Peacekeeping Group was mostly composed of infantry, but it also had an engineer subunit and a medical subunit. The Haiti Reconstruction Support Group and the South Sudan Reconstruction Support Group were mainly made up of engineers but also had an infantry subunit and a medical subunit. What makes the recent change in organizational structure more significant is how all of the functional elements work together.

This study identifies two types of mission formation. The first is *a mechanical mission formation*, where repressive norms and sanctions bind individuals. The mechanical mission formation is likely to be found in a unit where one branch plays a dominant role and the others a supporting role. According to an interview with an officer who participated in the UNPKO missions in East Timor, Iraq and Lebanon (on 11 November 2018, at the camp of the Lebanon Peacekeeping Group in Lebanon), there was discord between individuals from the different army branches in UNPKO units until the late 2000s. Mistrust and conflict between engineer officers and infantry officers were severe in the 198th Engineer Battalion

Table 7.1 South Korea's peace operation units (1991–present)

Type	From	To	Country	Name of unit	Size	Per capita budget (US$1,000)
MNFPO	1991	1991	Saudi Arabia	Gulf War Medical Assistance Group	134	64.9
	2001	2007	Afghanistan	924th Medical Assistance Group	100	131
	2003	2007	Afghanistan	100th Construction Engineer Group	150	114
	2003	2008	Iraq	320th Medical Assistance Group	100	88
	2003	2008	Iraq	1100th Construction Engineer Group	600	38.3
	2004	2008	Iraq	Zaytun Peace and Reconstruction Division	3,600	56.4
	2009	Present	Somalia	Somali Coast Convoy Squadron	310	109.3
	2010	2014	Afghanistan	Afghanistan Reconstruction Support Group	350	110.8
				Mean	668	89.1
UNPKO	1993	1994	Somalia	189th Engineer Battalion	250	42.4
	1994	2006	West Sahara	W. Sahara Medical Assistance Group	50	52
	1995	1996	Angola	101st Field Engineer Group	198	66.6
	1999	2003	East Timor	522nd Peacekeeping Group	419	22.9
	2007	Present	Lebanon	Lebanon Peacekeeping Group	350	41.4
	2010	2012	Haiti	Haiti Reconstruction Support Group	250	52
	2012	Present	South Sudan	S. Sudan Reconstruction Support Group	300	36
				Mean	259.5	44.7
DEC	2010	Present	UAE	UAE Military Training Cooperation Group	150	80.6
	2013	2014	Philippines	Joint Support Group for the Philippines	540	52.4
				Mean	345	66.5

Source: National Assembly Dispatch Agreement for each unit.
Note: size and budget = strength and budget for the first year.

Table 7.2 Composition of South Korea's UNPKO units

Name of unit	HQ (%)	Subunit			Total (%)
		Engineer (%)	Medical (%)	Infantry (%)	
189th Engineer Battalion	36.4	60.8	2.8	0.0	100
W. Sahara Medical Assistance Group	52.4	0.0	47.6	0.0	100
101st Field Engineer Group	20.8	64.6	3.0	11.6	100
522nd Peacekeeping Group	52.1	NA	NA	47.9	100
Lebanon Peacekeeping Group	52.4	3.9	3.6	40.1	100
Haiti Reconstruction Support Group	26.3	50.0	6.7	17.1	100
S. Sudan Reconstruction Support Group	3.0	70.0	7.0	20.0	100

Source: The National Assembly Dispatch Agreement for each unit; Republic of Korea Joint Chiefs of Staff, 1998.

in Somalia, the first UNPKO unit in South Korea's history, where the commander was a lieutenant-colonel engineer officer, and the executive officer was an infantry officer. It was in part because engineer officers had no chance to command infantry officers in South Korea, where engineer officers play a supporting role for infantry commanders. This type of dissonance was common in Iraq, where the engineer branch and the medical branch played a dominant role in providing reconstruction support for local people.

In the late 2000s in Lebanon, this mechanical mission formation quickly changed to *an organic mission formation*, where complementarities between individuals form interdependence. In Lebanon, a sense of organic solidarity among individuals from different army branches was far more robust than other UNPKO units. For example, the lack of participation and communication often creates misunderstandings and complaints to medical officers in South Korea. Medical officers in an infantry battalion are usually lieutenants who have just finished their internship and thus usually cannot take part in the decision-making process as a battalion staff at all. Because the medical department rarely provides general support or direct support to an infantry battalion in the South Korean field army, even medical officers with a higher rank have difficulty in understanding how to work in the combined combat battalion. According to an interview for this study (on 13 October 2018, at the camp of the Lebanon Peacekeeping Group in Lebanon), however, the interaction between infantry officers and medical officers was very different in Lebanon for at least three reasons.

First, the Lebanon Peacekeeping Group is smaller than an infantry battalion but has a medical department equivalent to a division medical department. A commander of the medical department is a staff for medical affairs and participates in the decision-making process. Second, the medical department in Lebanon is made up of numerous medical specialists so that it can provide necessary medical support in a timely manner. Finally, the medical department in Lebanon cannot support local people without the cover of the infantry unit. Communication to prepare medical assistance and ensuing an understanding of each other could reinforce cooperation. In Lebanon, these cooperative relations can be easily found between infantries and engineers too. The wonder is, then, how the South Korean military could construct the institutional logics that contributed to the shift from mechanical mission formation to an organic mission formation.

Competitive isomorphism and organic mission formation

Unit task

We first examine if mission formations resemble one another when the unit task is similar. There are four unit tasks – medical support, construction support, humanitarian support and military enforcement – that the National Assembly dispatch agreement specified (for unit tasks of each peace operation unit see Appendix). First, medical support involves activities such as patient treatment and disease prevention. Second, construction support involves activities such as the construction of roads, bridges and public facilities. Third, humanitarian support involves material or logistic assistance to the victims of a natural or human-made disaster. Finally, military enforcement is an activity of infantry units and includes the use of forces for peacebuilding. Next, we use multidimensional scaling (MDS) to visualize the similarity of peace operation units in terms of the unit task. For this analysis, we construct Matrix$_{ij}$, where i and j are one of the peace operation units, and give one for two peace operation units that carry out the same task. In the MDS diagram, two peace operation units come closer if they carry out similar jobs.

Figure 7.1 shows the changing nature of unit tasks assigned to South Korea's MNFPO and UNPKO units. In the early stage in the 1990s, there was no difference in unit tasks between MNFPO and UNPKO. MNFPO units and UNPKO units provided either medical or construction support to multinational forces commanded by the US and the UN. South Korea's peace operation units began providing medical or construction assistance, together with humanitarian aid in 1999. In the most recent stage, the difference between MNFPO and UNPKO becomes clear. UNPKO units have moved toward providing local people with more integrated reconstruction assistance, whereas MNFPO units have been more specialized in providing military enforcement.

If the unit task determined the mission formation, the two nearest peace operation units in Figure 7.1 should share the same mission formation. However, the similarity in a unit task cannot consistently account for the difference in mission formation among peace operation units. The UNPKO units in East Timor (A in Figure 7.1) and Lebanon (B in Figure 7.1) carried out the same unit task, providing military enforcement and humanitarian assistance to restore peace and stability in East Timor and Lebanon, respectively. In East Timor, infantry played a critical role, and the others a supporting role. The primary mission of engineer

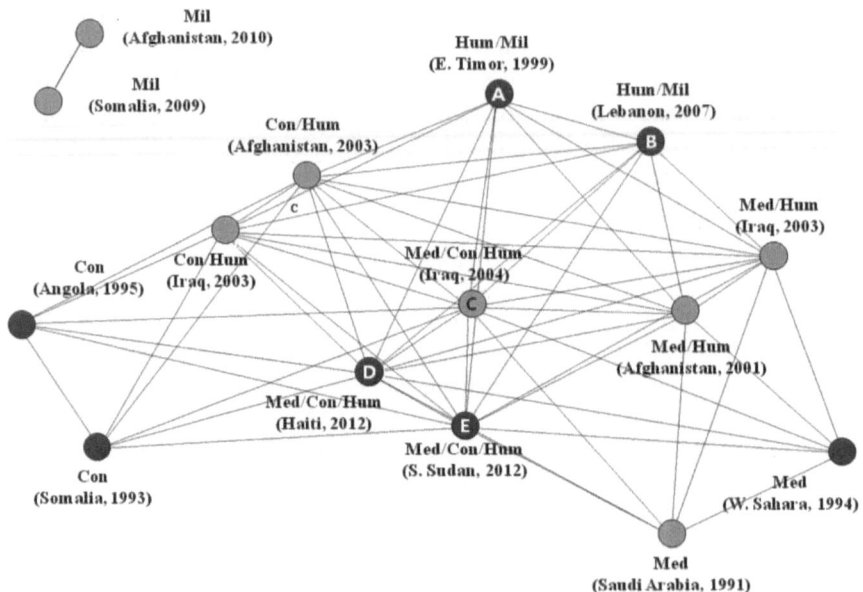

Note: Med = medical support; Con = construction support; Hum = humanitarian support; Mil = military enforcement

● = MNFPO; ● = UNPKO.

Figure 7.1 The network of overseas military units with similar unit tasks

officers and medical officers in the 522nd Peacekeeping Group was not to engage in humanitarian relief for local people but to directly support the infantry subunits (Song, 1999). On the contrary, this mechanical mission formation in East Timor was different from the organic mission formation that first appeared in the Lebanon Peacekeeping Group. Since then, it repeatedly appeared in the Haiti Reconstruction Support Group and the South Sudan Reconstruction Support Group.

Operational environment

From a competitive isomorphism perspective, "the diversity of organizational forms is isomorphic to the diversity of environments" (Hannan and Freeman, 1977, p. 939). Thus, the differences in the operational environment can explain the differences in mission formation. According to an interview for this study (on 11 November 2018, at the camp of the Lebanon Peacekeeping Group in Lebanon), political stability seems to be decisive in determining mission formation. In South Korea, where North Korea's war threat always exists, infantry officers necessarily play a dominant role. In Lebanon, where the political situation is relatively stable, there is no need to engage in combat activities. Instead, officers from combat service branches and combat service support branches have more to do than infantry officers. Then, it is easy to conclude that functional elements in a peace operation unit will be more dependent on each other when political stability is higher. However, this is not the case, as Figure 7.2 shows.

Although Haiti and South Sudan are politically less stable than East Timor, the Haiti Reconstruction Group and the South Sudan Reconstruction Support Group have a more cooperative atmosphere than the 522nd Peacekeeping Group in East Timor. It is also still difficult to explain why there was no such collaborative atmosphere in Angola, where political stability was similar to Lebanon. According to Chiara Ruffa (2014, p. 199), peacekeepers "construct the operational environment differently in a way that is consistent with their different military behavior". This perspective is not new, because previous literature on operational style in multinational cooperation has attributed variation in the role of soldiers to contextual, societal and political determinants. Ruffa contributed to the current debate by empirically examining how the interpretation of the operational environment influenced the way French, Ghanaian, Italian and South Korean peacekeepers behaved in the United Nations Interim Force in Lebanon (UNIFIL). Ruffa also argues for the importance of the experience that makes sense of an operational environment. From this perspective, Ruffa contends that South Korean peacekeepers, who were new to peacekeeping, displayed high force protection measures and a strong focus on Civil-Military Coordination (CIMIC) because they perceived a high threat level, a well-defined enemy and a strong commitment to the PKO mission in Lebanon.

However, Ruffa's argument has at least two flaws. First, the South Korean history of the UNPKO mission began on 31 July 1993, when 250 engineers arrived in Somalia to participate in the United Nations Operation in Somalia II. South Korea

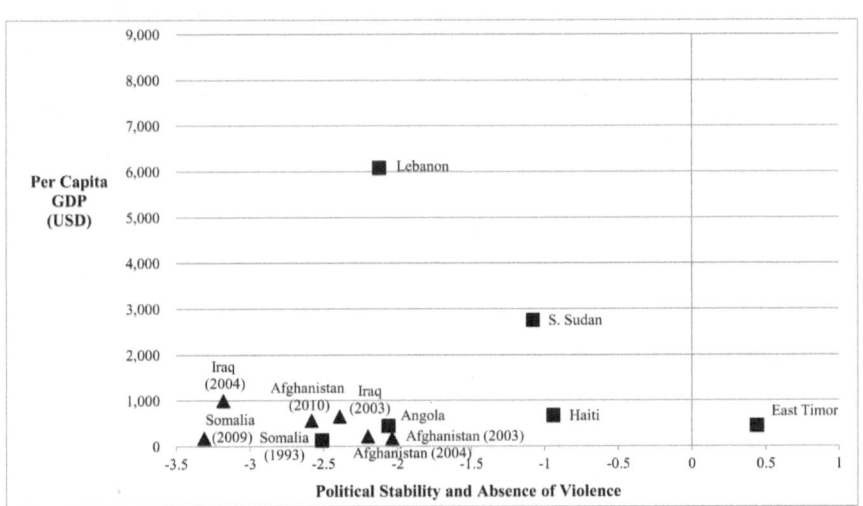

Note: ▲=MNFPO ■=UNPKO

Source: per capita GDP = The UN Statistics Division; Political Stability and Absence of Violence = World Governance Index

Figure 7.2 Distribution of peace operation units by political and economic condition

had more than 14 years of UNPKO experience when it dispatched peacekeepers to the UNIFIL. What South Korean peacekeepers thought and did in Lebanon reflect not only their interpretation of the operational environment in Lebanon but also their experiences in the past UNPKO missions. Second, high force protection measures and ongoing CIMIC activities are a common feature of all South Korean UNPKO units, irrespective of the operational environment. South Korea's intense concern for force protection and CIMIC may be attributed to national instruction because peacekeepers act on instructions from the Joint Chiefs of Staff (JCS) unless those instructions hinder the implementation of the UN mission's mandate (UN DPKO, 2008). In East Timor, Iraq and Lebanon, where political and economic situations varied greatly, South Korean peace operation units have carried out independent CIMICs called Blue Angel, Green Angel, and Peace Wave, respectively.

Institutional isomorphic change in UNPKO mission formation

The emergence of institutional environment

From an institutional isomorphism perspective, organizations tend to seek legitimacy rather than efficiency (DiMaggio & Powell, 1991, p. 65). Organizations gain legitimacy by acting in ways that are consistent with values and norms embedded in an institutional environment. Once we adopt an institutional isomorphism perspective, we surely have to ask how an institutional climate emerges in a different form from the past. According to DiMaggio and Powell (1991, p. 65), the structure of the organizational field is not determined a priori but is institutionally defined.

In the 1990s, South Korea's peace operation units began constructing an actual organizational field, where they "partake[s] of a common meaning system and whose participants interact more frequently and fatefully with one another than with actors outside the field" (Scott, 1995, p. 56), as Table 7.3 shows. It was for at least two reasons. First, they began working closely with other peace operation units in different countries. In this cooperative process, a typical pattern of a coalition where diverse functional elements work in conjunction with one another has emerged.

Table 7.3 Distribution of South Korea's peace operation by year and country

Country	Year				
	1991–1995	*1996–2000*	*2001–2005*	*2006–2010*	*2011–2015*
South Sudan					
Haiti					
Lebanon					
Iraq					
Afghanistan					
East Timor					
Angola					
W. Sahara					
Somalia					
Saudi Arabia					

Second, people who have participated in one peace operation often participate in another peace operation and thus share their know-how with new peacekeepers. According to an interview for this study (on 11 November 2018, at the camp of the Lebanon Peacekeeping Group in Lebanon), 71 of 331 (21.5 percent) South Korean peacekeepers in Lebanon participated in peace operations more than once.

DiMaggio and Powell (1991, pp. 67–74) identify three mechanisms – mimetic isomorphism, normative isomorphism and coercive isomorphism – to explain how institutional isomorphism occurs. First, mimetic isomorphism results from voluntary imitation of practices adopted by other organizations. The mimetic isomorphism is most relevant when uncertainty about goals and the environment is high. Second, normative isomorphism is attributable to the spread of practices that professionals believe desirable. Third, coercive isomorphism occurs owing to the coercive imposition of formal and informal pressures by other organizations, such as the government.

Uncertainty about peace operation and mimetic isomorphism in the 1990s

When South Korea began participating in peace operations in the early 1990s, doubts about them prevailed for three reasons. First, there was concern about the negative impact of the deployment of combatants on the preparedness for North Korean threats. Second, there was no relevant guideline for peace operations. The Ministry of National Defense was able to establish directives concerning the UNPKO dispatch process in 1995, 4 years after the first peace operation ended. It was only in 2012 that the South Korean Army published a field manual for peace operations for the first time. Third, the South Korean people were increasingly worried about the safety of soldiers. According to a national survey about peace operations abroad in 2015, 79.9 percent of South Korean people were still concerned that soldiers could die (Republic of Korea Ministry of National Defense, 2015).

Owing to increasing uncertainty in the 1990s, experiences of the Vietnam War (1964–1973) had a significant impact on mission formation of peace operation units. At the breakout of the Vietnam War in the earlier 1960s, South Korea wanted to send troops to Vietnam on the condition of obtaining US military aid for the modernization of the South Korean military. During the Vietnam War, therefore, it was the US government that decided which units should go to the Vietnam War (1964–1973). In May 1964, South Korea sent one Mobile Army Surgical Hospital (MASH). One medical company of size 130 and ten Taekwondo (Korean traditional martial art) instructors followed them. Soon after the Gulf of Tonkin incident broke out in August 1964, it sent a rear support unit consisting of noncombatants. One construction support group of size about 3,000 engineers, called "Pigeon Unit", followed in March 1965. As the ground battle was heating up in various parts of South Vietnam, at the request of the US, it reinforced one marine brigade (the 2nd Marine Brigade) and two army infantry divisions (the Capital Division and the 9th Division) sequentially until October 1966.

The organizational characteristics of the overseas military units, created between 1991 and 2004, showed the most dramatic instances of imitative modeling. South

Korea's PKO units were modeled on the old Vietnam War model that was familiar to senior military officers for two reasons. First, South Korea created MNFPO units and UNPKO units along the three branch lines – medical, engineer and infantry unit – like troops dispatched to Vietnam. Second, each MNFPO unit and UNPKO unit carried out its missions independently, even in the same country. For example, at the outbreak of the War on Terror in 2001, the South Korean government organized the 924th MAG and the 100th CEG and made them carry out a separate mission in Afghanistan. This pattern was reiterated in Iraq again. The South Korean government organized the 320th MAG and the 1100th CEG separately, and they conducted independent tasks in Iraq. As the US government officially requested the South Korean government to reinforce one more infantry division in Iraq in September 2003, the Iraq Peace and Reconstruction Division, called the Zaytun, which means olive in Iraq, was created in 2004 to integrate all the previously dispatched units.

Looking at the historical path along which South Korea's peace operation units have advanced, we can identify a pattern that cannot be explained by mimetic isomorphism. Although the mimetic isomorphism perspective is useful to explain the continuity of organizational forms until the 1990s, it cannot account for the change from a mechanical mission formation to an organic mission formation in the late 2000s, as Table 7.4 shows.

Unit requirement by the United Nations and coercive isomorphism

When South Korea is requested to participate in a peace operation, the JCS decides the operational concept and proposes the organizational structure of a peace operation unit to the Ministry of National Defense. According to Article 6 of the UNPKO Participation Act, the Ministry of National Defense submits overseas dispatch plans, including the contingent's mission and measures to ensure the safety of the troops, to the National Assembly. Considering whether the purpose of a peace operation is congruent with Article 5 of the Constitution, prescribing "The Republic of Korea shall endeavor to maintain international peace and shall renounce all aggressive wars", the National Assembly decides whether or not to agree to the overseas dispatch plan. As the South Korean government has to get consent from the National Assembly, it has to avoid the suspicion that peace operation units would be used differently from the UN mandate. For example, when South Korea first took part in UNPKO in 1993, the Ministry of Defense instructed that the number of engineers in the 189th Engineer Battalion should meet the requirement set out by the United Nations (Republic of Korea JCS, 1998, p. 62).

In the early 2000s, the United Nations developed a new principle called Responsibility to Protect (R2P), which "embodies a political commitment to end the worst forms of violence and persecution" (UN Office on Genocide Prevention and Responsibility to Protect, 2020). Accepting the principle of R2P as an international norm, the militaries of the industrial democracies began increasingly engaging in peace enforcement operations. Accordingly, the unit requirement by the United Nations could change consistently. If coercive isomorphism was

Table 7.4 Mimetic isomorphism between South Korea's overseas military units

Year	Overseas military unit		
	Medical unit	*Engineer unit*	*Infantry (navy) unit*
The 1960s	MASH	Pigeon Unit	Infantry and Marine Divisions
The 1990s	Gulf War Medical Assistance Group (Saudi Arabia, 1991) W. Sahara Medical Assistance Group (West Sahara, 1994)	189th Engineer Battalion (Somalia, 1993) 101st Field Engineer Group (Angola, 1995)	522nd Peacekeeping Group (East Timor, 1999)
The early 2000s	924th Medical Assistance Group (Afghanistan, 2001) Zaytun Reconstruction Support Division (Iraq, 2004) 320th Medical Assistance Group (Iraq, 2003)	100th Construction Engineer Group (Afghanistan, 2003) 1100th Construction Engineer Group (Iraq, 2003)	11th and 12th Civil Affair Brigade (Iraq, 2004)
The late 2000s	*Lebanon Peacekeeping Group (Lebanon, 2007)* *Haiti Reconstruction Support Group (Haiti, 2010)* *S. Sudan Reconstruction Support Group (South Sudan, 2012)*		Somali Coast Convoy Squadron (Navy) (Somalia, 2009) Afghanistan Reconstruction Support Group (Afghanistan, 2010)

Note: = □ MNFPO; ■ = UNPKO; italics = an organic mission formation; Dark area means that the two types of units were combined.

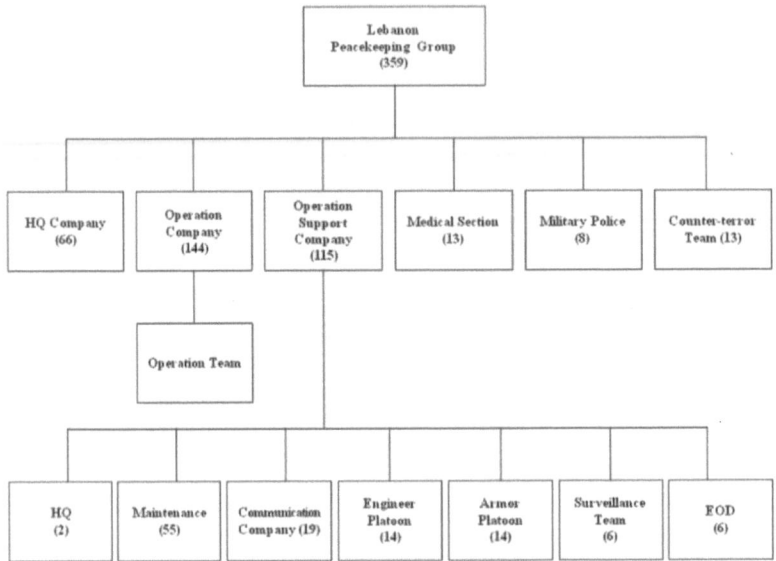

Figure 7.3 Organizational structure of the Lebanon Peacekeeping Group. Source: National
Assembly Agreement on the Extention of Overseas Dispatch for the Lebanon
Peackeeping Group (June 2008), p. 7.

dominant, South Korea's UNPKO units should have an organizational structure
that the UN specified. The organizational structure of the Lebanon Peacekeeping
Group is almost the same as the unit requirement by the UN. Most importantly,
there is little difference in the number of personnel in the medical and engineer subunits between the United Nations' unit requirement and the Lebanon
Peacekeeping Group, as Figure 7.3 shows (for the unit requirement for Light
Infantry Task Force in UNIFIL by the UN; see the UN DPKO, 2019, p. 11). Even
though coercive isomorphism is useful to explain the change to an integrated
organizational structure in the 2000s, it cannot answer the fundamental question
of this study: how could the South Korea peacekeepers build a cooperative relationship that they have rarely experienced before?

Professionalization and normative isomorphism

Institutions "operate primarily by affecting a person's perspective bets about the
collective environment and collective activity" (Jepperson, 1991, p. 147). From
this perspective, change from a mechanical mission formation to an organic
mission formation in UNPKO units can be attributed to a change in perception
about what activities are desirable for South Korea's UNPKO. According to
an interview for this study (on 11 November 2018, at the camp of the Lebanon
Peacekeeping Group in Lebanon), the interaction between an operation unit consisting of infantries and other operation support units is significantly different

from what they experienced in South Korea. In Lebanon, the collective identity as a member of the Lebanon Peacekeeping Group took precedence over the particular branch identity. It was through repeated discussion in operational meetings and accumulated experiences in the field, rather than the commander's instruction or repressive rule that peacekeepers could be aware of the importance of cooperation between different functional elements in overseas military units. This group identity as a member of a cooperative entity did not disappear, because predecessors in Lebanon shared their know-how with new peacekeepers.

In the 2000s, at least three conditions contributed to the expansion of organic mission formation. First, the Zaytun Peace and Reconstruction Division introduced the new concept of reconstruction assistance to South Korea's overseas military mission. Contrary to a previous peace operation unit whose fundamental purpose was to provide medical aid and engineering assistance to the multinational forces led by the UN or the US, the Zaytun Peace and Reconstruction Division provided reconstruction assistance to local people in Erbil, Iraq. This practice strengthened the legitimacy of peace operations so that the Lebanon Peacekeeping Group could actively engage in humanitarian aid, including various educational programs such as computer, baking and heavy vehicle driving skills.

Second, the Zaytun Peace and Reconstruction Division produced a large number of experienced middle-ranking officers who could successfully serve in peace operation units. Those who once served in the Zaytun Peace and Reconstruction Division could make great use of their know-how about reconstruction assistance for other peace operations in Lebanon, Haiti and South Sudan. One more important step forward for South Korea's UNPKOs was the establishment of the International Peace Support Group (IPSG). About 1,000 officers in the IPSG are learning basic tactical tasks for UNPKOs. The IPSG provides a public sphere where peacekeepers can share knowledge and experience. According to an interview with a staff of the Lebanon Peacekeeping Group (on 11 November 2018, at the camp of the Lebanon Peacekeeping Group in Lebanon), all staff had frequent contact and exchanges with each other when they stayed in the IPSG.

Finally, along with the expanded number of officers who have experienced peace operations, academic research on peace operations has increased simultaneously, as Figure 7.4 shows. In the 1990s, the South Korean military first began to educate officers on how to conduct peace operations. The Joint Forces Staff College established the Department of PKO Studies in 1995, and the National Defense University followed this effort with the establishment of the PKO Center in 2004. Integrating the Department of PKO studies, the PKO Center developed into the most extensive professional research and education institution, consisting of 29 instructors and staff in 2010. In the early 2000s, organizational norms and shared knowledge from Iraq and other peace operations could create the social pressure that had South Korean peacekeepers get used to an organic mission formation.

Conclusion and discussion

This study examined how a mechanical mission formation changed into an organic mission formation in South Korea's UNPKO units. Three mechanisms of

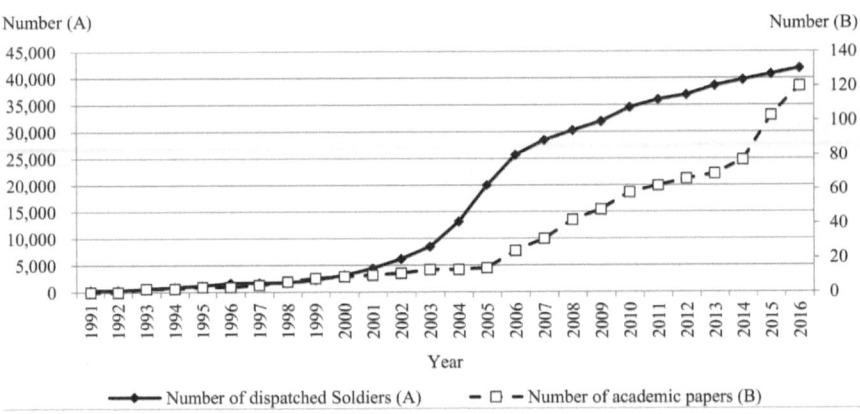

Figure 7.4 Cumulative number of peace operation participants and academic journal papers about peace operation

isomorphism account for the evolution of mission formation in peace operation units. First, mimetic isomorphism created all peace operation units in the same form as the Vietnam War model. In the earlier period, between 1991 and 2004, the peace operation units were organized separately along army branches and conducted missions independently. Second, coercive isomorphism produced a homogeneous organizational structure in the early 2000s because the unit requirement of the United Nations changed in a way that integrates various functional elements. Finally, in the late 2000s, normative isomorphism created social pressure that made all the individuals from different army branches dependent on each other, thanks to the rapid increase in PKO experts and shared knowledge in academia.

The approach used in this study is not new to military organizational research (Farrell and Terriff, 2002). However, the above findings have a few implications for the nature of mission formation. First, mission formation is not a product of decisions by a commander of the UNPKO unit, but the commander's discourse is essential to understand mission formation. All interviewees commonly stated that commanders play a critical role in creating a cooperative atmosphere in UNPKO units. Commanders have a significant influence on the attitude of the staff from the different army branches toward each other because they can put forward their vision of UNPKO and templates for action. More specifically, words that a commander uses in everyday life, rather than a specific instruction, are decisive in producing a more cooperative atmosphere. Second, identity and norms are central to understanding mission formation. The collective identity as a peacekeeper helps explain the way South Korean peacekeepers have acted in conflict areas with some regularity. To better understand mission formation, we need to empirically examine social interaction among peacekeepers contributing to creating and maintaining their collective identity.

Appendix

Table 7.A.1 Unit task of South Korea's overseas military unit (1991–present)

Type	Name of unit	Medical support	Construction support	Humanitarian support	Military enforcement
MNFPO	Gulf War Medical Assistance Group	○			
	924th Medical Assistance Group	○			
	100th Construction Engineer Group		○	○	
	320th Medical Assistance Group	○			
	1100th Construction Engineer Group		○	○	
	Zaytun Peace and Reconstruction Division	○	○	○	
	Somali Coast Convoy Squadron (Navy)				○
	Afghanistan Reconstruction Support Group				○
UNPKO	189th Engineer Battalion		○		
	W. Sahara Medical Assistance Group	○			
	101st Field Engineer Group		○		
	522nd Peacekeeping Group				○
	Lebanon Peacekeeping Group			○	○
	Haiti Reconstruction Support Group	○	○	○	
	S. Sudan Reconstruction Support Group	○	○	○	

Source: Republic of Korea National Assembly, *National Assembly Agreement on Overseas Dispatch* (various issues), Retrieved from http://www.mofa.go.kr.

References

Article 5 of the Constitution. Retrieved from https://elaw.klri.re.kr/kor_service/lawView.do?hseq=1&lang=ENG

Article 6 of the United Nations Peacekeeping Operations Participation Act. Retrieved from https://elaw.klri.re.kr/kor_service/lawView.do?hseq=33674&lang=ENG

Clausewitz, C. (2008). M. Howard & P. Paret (Eds. and Trans). *On War*. NJ: Princeton University Press.

DiMaggio, P.J., & Powell, W.W. (1991). The iron cage revisited: Institutional isomorphism and collective rationality in organizational field. In W.W. Powell & P.J. DiMaggio (Eds.), *The new institutionalism in organizational analysis* (pp. 63–82). Chicago, IL: University of Chicago Press.

Farrell, T., & Terriff, T. (2002). *The sources of military change: Culture, politics, and technology*. Boulder, CO: Lynne Rienner.

Hannan, M.T., & Freeman, J.H. (1977). The population ecology of organizations. *American Journal of Sociology*, *92*(5), 929–964.

Jepperson, R.L. (1991). Institutions, institutional effects, and institutionalism. In W.W. Powell & P.J. DiMaggio (Eds.), *The new institutionalism in organizational analysis* (pp. 143–163). Chicago, IL: University of Chicago Press.

Kaldor, M. (1990). *New and old wars: Organized violence in a global era*. Cambridge: Polity Press.

Orrù, M., Biggart, N.W., & Hamiton, GG. (1991). Organizational isomorphism in East Asia. In W.W. Powell & P.J. DiMaggio (Eds.), *The new institutionalism in organizational analysis* (pp. 361–389). Chicago, IL: University of Chicago Press

Republic of Korea Army. (2009). *Peace operation*. Seoul: Republic of Korea Army.

Republic of Korea Joint Chiefs of Staff. (1998). *History of Republic of Korea Army PKO*. Seoul: JCS.

Republic of Korea Ministry of National Defense. (2015). Dandelion Letter (Korean). Retrieved from www.mnd.go.kr/user/mnd/upload/pblictn/PBLICTNEBOOK_2015 11250217393780.pdf

Republic of Korea Ministry of National Defense. (2016). *Defense White Paper*. Seoul: Ministry of National Defense.

Ruffa, C. (2014). What peacekeepers think and do: An exploratory study of French, Ghanaian, Italian, and South Korean armies in the United Nations interim force in Lebanon. *Armed Forces and Society*, *40*(2), 199–225.

Scott, W.R. (1995). *Institutions and organizations*. Thousand Oaks, CA: Sage.

Song, Y. (1999, October 11). South Korea's PKO. *Weekly Defense Review 787*.

The United Nations Department of Peacekeeping Operations. (2019, February). Statement of unit requirement for light infantry task force in UNIFIL.

Tilly, C. (1992). *Coercion, capital, and European states, AD 990–1992*. Cambridge, MA: Blackwell.

The United Nations Department of Peacekeeping Operations/Department of Field Support. (2008). Authority, Command, and Control in United Nations Peacekeeping Operations, Article 51.

The United Nations Office on Genocide Prevention and Responsibility to Protect. (2020). Responsibility to protect. Retrieved from www.un.org/en/genocideprevention/about-responsibility-to-protect.shtml

Vlahos, M. (2003). *Perspectives on military transformation towards a global security force*. Johns Hopkins University Applied Physic Laboratory. Retrieved from www.jhuap.edu/ourwork/nsa/papers/PerspectivesMilitaryTransformation.pdf

8 "Democracy… 120 mm at a time"

Mission formations and operational entrapments in post-9/11 Afghanistan

Thomas Randrup Pedersen

Introduction: Ugly

Two British Apache helicopters fly roaring in low above us, each armed with a chain gun, rockets and Hellfire missiles. So much fire power. So impressive. So terrifying. No wonder the attack helicopters go by the call sign "Ugly", let alone have become known locally in Afghanistan as "the monsters" (Matthews, 2016). I am in the back of an infantry mobility vehicle, a Cougar MRAP,[1] leased from the US marines in the adjacent Camp Leatherneck. Now, the heavy armored personnel carrier forms part of the Danish force protection section Fenrir. The 22-tonne vehicle is designed for protection against small arms fire, rocket-propelled grenades, mines and, not least, improvised explosive devices. In short, the Cougar is designed for the asymmetric warfare characterizing the post-9/11 wars in the early 21st century – the "New Wars" (Kaldor, 2012) or the "Hybrid Wars" (Hoffmann, 2007) – and we feel safe.

Fenrir's four-vehicle convoy makes its way through the flight line area at the outer perimeter of Camp Bastion – the UK-operated mother base in the "Desert of Death", Helmand, Afghanistan. The long-standing NATO-led mission in the country, the International Security Assistance Force (ISAF), is drawing to a close. There are, however, still noisy Apaches and other military aircraft in the sky over the huge base complex each and every day. To see and hear two of the Apaches this close is quite something. It brings us closer to the war machine of the US-led coalition force that still deals in death and destruction "outside the wire". It brings us closer to where the action is.

In the Cougar's front seats, vehicle commander Private Nordrup and his driver Private Thorsminde both cheer on the two flying gunships with great enthusiasm.[2] In the back seat, I fall silent, struggling with mixed emotions. I feel fascinated yet repelled. That is not the first time, nor will it be the last, during my fieldwork as an anthropologist "embedded" in the Danish ISAF force (Pedersen, 2017b, 2019c). For instance, feelings of ambivalence once again make themselves present when I follow the Danish tactical air control party (TACP) on a static display visit to the British Army Air Corps's 654 Squadron in Bastion. This time, the mixed emotions are evoked first by watching Apache "kill TV",[3] and then by shooting new social media pictures of my fellow countrymen posing in front of one of the lethal "monster machines". What is in a kill footage? Professionalism? Potency? A morale boost? The promotion of democracy? What is in a (self-) image? Adventurousness? Manliness? The potential of violence? Bonds between nations and militaries, people and technologies?

There are bound to be numerous answers to these questions. In this chapter, however, I emphasize just one: the desire for action. It is an answer that somewhat resonates with the one given by the sociologist Anthony King (2010) in his study of the British Helmand campaign. Commanders in the British armed forces, King argues, were mounting operations out of a desire to act. Such desire, King (2010, p. 322) notes, "seems to be a product of the ethos of the British military and the professional expectations of the British officer corps". After all, as King (2010, p. 323) points out, British defense doctrine elevates combat into the ultimate professional ideal by way of singling out a "warfighting ethos" as one of the founding principles for the "British way of war". Accordingly, "Officers must be decisive. They must constantly seek to seize the initiative through offensive action" (King, 2010, p. 323).

To be sure, King (2010) is on to something, and the British case does, to a certain extent, resemble its Danish counterpart. At least, that is so in the case of my own research with low-ranking officers and soldiers in the Danish ISAF force. Rather than seeking to step into character as the "hearts and minds"-winning "soldier-diplomat", the tankers and the recce men, the dragoons and the hussars I have followed (Pedersen, 2017a, 2017b, 2019a, 2019b) aspired above all to get into real combat and thereby to become warfighting "soldier-warriors". As such, my interlocutors in the tank platoon Loki and in the force protection section Fenrir were not so much concerned with embodying a force for good as with embodying a force of death and destruction, that is, they were not so much concerned with making the world into a "better place" as with making themselves into "greater warriors". As one of Loki's gunners wrote on the rim of his tank gun's yellow muzzle cover: "Democracy … 120 mm at a time". In keeping with tradition, the gunner had incidentally also painted a bold, happy smiley on the front of the very same muzzle cover. Happiness, it seems, was a warm gun.

In the present inquiry, I conceive the desire for action as an existential driving force mediated through relationships with others, humans and things. In the following pages, I ethnographically investigate a range of experiential spheres in which such self-motivating yearning is intimately tied to the temporary, multinational, tailor-made mission formations defining coalition warfare in the post-9/11 years. The aim of this chapter is twofold. First, I seek to contribute empirically, methodologically and theoretically to what Eyal Ben-Ari, Uzi Ben-Shalom, Thomas Brønd and Carmit Padan describe as the emergent "sociology of mission formations" (Ben-Ari et al., this volume). Second, by the same token, I endeavor to make an anthropological contribution to the reinvigoration of the research on classic subjects in military sociology, namely motivation and combat.

Empirically, I pursue the dual aim of this inquiry by drawing upon the ethnographic fieldwork material that I have co-produced with Danish "grunts" toward the end of "ISAF's war" in Afghanistan.[4] More specifically, the analytical sections of this chapter rest upon my fieldwork with the tank platoon Loki. In addition, as a way of concluding the chapter, I put my analysis into perspective by way of reconnecting with the force protection section Fenrir, introduced above. Loki and Fenrir were both constituent elements of alternating mission formations, combined for the specific ISAF mission, or even for specific ISAF operations. Both units were 16 strong and formed part of the same 6-month rotation team, the same provisional Danish ISAF contingent (DANCON/ISAF). Fenrir was formed by 1st Light Recce Squadron, 3rd Recce Battalion, the Guard Hussar Regiment. It was a combat unit designed for covering and escorting Danish and allied forces.[5] By comparison, Loki was stood up by 2nd Tank Squadron, 1st Armoured Battalion, the Jutland Dragoon Regiment. In theater, Loki was in charge of a multi-element tank detachment consisting of various "enablers" and "assets", such as the TACP mentioned above. The up to 60-strong detachment was designed for advanced combat support of allied ground forces. It was recurrently seconded to multinational task forces and worked closely together with British, Estonian and US coalition partners.

Methodologically, I have generated empirical data on different mission formations through multi-sited fieldwork, periodically coming and going over the course of 12 consecutive months (Pedersen, 2017b). I have followed and interviewed the same individual officers and soldiers before, during and after deployment to Helmand (ibid.). Regrettably, as often seems to be the case with social scientists studying the armed forces (Ben-Shalom, this volume), I was not allowed physical access to operations in theater. Consequently, this chapter is primarily based on post-operation interviews, several of which were conducted relatively close to the narrated events and experiences in time and space. Additionally, this chapter does, to a lesser extent, rest on my participant observation of everyday camp life, including after-action review sessions. On this basis, I seek to study mission formations and military action in "a view from within". I take inspiration from phenomenological anthropology (Jackson, 1996, 1998; Desjarlais & Throop, 2011; Mattingly, 2012, 2014; Ram & Houston, 2015) and adopt a first-person perspective to convey insights into the individual lifeworlds of my interlocutors, their desires for action, their experiences of mission formations and their reflections on the (possible) use of military violence.

Theoretically, I turn to Michael D. Jackson's existential anthropology to conceptualize the desire for action as an expression of the human struggle for Being; for reclaiming a sense of agency in the world (Jackson, 1998, 2005, 2013) and, as I add in regard to this study's military context, for reclaiming a sense of "existential potency" within the bounds of one's military professionalism: one's military ethics, knowledge, skills and competencies. Furthermore, I draw upon Alberto Corsín Jiménez and Chloe Nahum-Claudel's anthropology of traps in order to conceptualize mission formations as trap configurations/entrapment processes (Jiménez & Nahum-Claudel, 2019). My theoretical ambition is to contribute to what the authors of the introduction to this volume refer to as a "move from a sociology of combat units to a sociology of mission formations" (Ben-Ari et al., this volume; see also Ben-Ari et al., 2010; Ben-Ari, 2018), that is, a move from the classic military sociology's study of "textbook units", such as platoons, companies and battalions, to the new sociology of mission formation's study of "tactical assemblages" of mission-specific "plug-and-play" forces in theater, such as detachments, battlegroups and task forces.

This move toward a new sociology of mission formations suggests a shift of emphasis from a focus on the cohesion and effectiveness of combat units in conventional inter-state warfare to new lines of inquiry critically questioning specific mission formations' social organization of violence and accomplishment of collective action in the wars of our time: the New Wars or the Hybrid Wars (Ben-Ari et al., this volume). In this vein, the present inquiry scrutinizes experiential spheres in which the quest for existential potency affects and is affected by various ways in which mission formations collectively restrain or release military action in post-9/11 Afghanistan. I argue that the desire for action, what I call the "struggle for existential potency", becomes entangled and entrapped in mission formations charged with the social organization of violence on the ground. I further argue that mission formations harbor conflicting traps: some serve to restrict the use of military action, such as rules of engagement (ROE); others promote the resort to military violence, such as the competition for warriorhood among multinational coalition forces; and still others defeat their own purpose by way of restraining rather than releasing military action, such as different doctrinal and command practices. I show that frustrations and existential impotence tend to follow from military inaction, whereas the utilization of military violence, on the other hand, brings about a sense of existential potency and existential fulfillment.

The remainder of this chapter is organized into four main parts. First, I deepen my theorizations on the desire for action and contemporary mission formations. Second, I provide an ethnographic account of my interlocutors and the mission formations they were part of. Third, I offer an anthropological analysis of the intimate relationships between the desire for action, mission formations and military violence. Fourth, in conclusion, I underscore the study's key findings, and what I see as their implications for the emergent research agenda on mission formations in the early 21st century.

Mission motivations/formations: Theoretical elaborations

Existential potency

As a case of professional soldiers in early 21st century volunteer forces, my interlocutors were, contrary to citizen soldiers in the 20th century (King, 2013), primarily motivated neither by political goals nor by patriotism or comradeship. Rather, characteristically of what Charles Moskos and James Burk (2019[1994]; see also Moskos, 2000) have described as "postmodern militaries", and as an instance of what Fabrizio Battistelli (1997) has referred to as "post-modern motivation", the Danish troops I have followed were first and foremost driven by a "desire for adventure, new and meaningful personal experience" (ibid., p. 471).

More specifically, my interlocutors were driven by a Romantic quest for self-transformative war experiences (Pedersen, 2017a, 2017b, 2018, 2019a; see also Harari, 2008) – a quest for becoming a self in the image of the "true warrior"; a quest for self-discovery and self-development by way of trying one's strength on the world; a quest pursued through the search for military adventure and, ultimately, for military action.

By the same token, this quest for soldierly self-becoming constitutes a quest for finding the answers to the existential questions: Who am I? What stuff am I made of? Will I freeze in the face of combat? Will I flee? Will I fight? As such, this quest is one that strongly resonates with the citizen soldier's search for proving his manliness to himself and others through combat in the 20th century (King, 2013). However, combat today, it seems, does not so much amount to a *rite de passage* into manhood (ibid.) as a rite of passage into selfhood. What the professional soldier seeks to achieve through military violence in the early 21st century tends not so much to be articulated in terms of proving oneself as a "real man" as in terms of proving one's "real me", the existence of an "authentic self". At least this is so in the case of the Danish combat troops with whom I have conducted fieldwork (Pedersen, 2017a; 2017b; cf. Brænder & Andersen, 2013; Brænder, 2016; Lyk-Jensen & Glad, 2018; Lyk-Jensen & Pedersen, 2019).

Accordingly, in this study, I prefer to conceptualize the desire for action in existential terms rather than regarding it as a question of performing masculinity and searching for manhood. Calling Michael D. Jackson's existential anthropology into play, I conceive the desire for action as an expression of what Jackson (2005, p. 143) has described as the human struggle for striking some sort of balance between "being an actor and being acted upon". This continuous struggle to "strike a balance between autonomy and anonymity" (Jackson, 2013, p. 133) is a driving force in human life, an existential imperative: "Every person demands, as a condition of being human, that he or she has some say over his or her own existence, some place in the world where his or her actions count" (ibid.). Along this line of thinking, I suggest that the desire for action and the use of military violence – against powerful or anonymous combatants – can be understood, with Jackson (1998, p. 76), as a desire for and a claim to the restoration of "a sense of being able to act, of possessing presence and substance, of being able to have an effect on the world that seems to have rubbed one out or cast one down". As such, I contend that the desire for action and the use of military violence form integral parts of the human struggle for what I coin "existential potency", that is, for the power of being able to affect the world (the specific battlespace) and reclaim a sense of agency that augments one's sense of (military) presence and significance, let alone boosts one's sense of self (as "soldier-warrior") (cf. Jackson, 2007).

Traps/entrapments

In their anthropology of traps, Alberto Corsín Jiménez and Chloe Nahum-Claudel (2019, p. 384) describe the notion of the trap as "trap/entrapment", that is, as a notion with a duplex identity as a noun and as a verb. As a noun, traps refer to material designs with lethal potentials (ibid.). As a verb, on the other hand, to trap – as I conceive it here – refers to embodied processes of entrapping, of tracking down and encircling. Both as a noun and as a verb, trap connects and entangles trappers and prey just as much as it joins and enmeshes materiality and mind, technologies and landscapes (ibid.). In this way, we can think of mission formations as traps/entrapments: as material designs, mission formations constitute concrete configurations of combat and support troops, all tailor-made to trap specific kinds of "prey" – "enemies", that is – on the tactical level in theater. On the other hand, as embodied processes of entrapping "the enemy", mission formations make up dynamic procedures for operating in the battlespace, such as standing or standard operating procedures (SOPs) for chains of commands, for military communications, for tactical maneuvering, for opening fire, etc. By implication, mission formations, I suggest, potentially harbor internal cultural dynamics in the form of inter-force competition for "setting up" the best traps and thus delivering the best performances in the battlespace, performances measured against military professionalism and existential potency, against the letter of the SOPs on the one hand and, on the other, against the ethos of "the warrior".

Based on my fieldwork with Danish ISAF troops, I identify three constituent elements in specific mission formations: doctrines/commanders, rules/systems and identity/time. I posit that each of these three two-sided elements makes up an existential trap as a part of the given mission formation, which in itself forms a material/embodied trap as a whole. Each of the three dual traps entangle and entrap the desire for action and the actual use of military violence in specific mission formations. Accordingly, each trap plays a crucial part not only in the mission formation's collective action and for the efficacy of its entrapment process, the actual engagement of "the target" but also for the resilience of its individual "trappers", the officers and soldiers who make up the mission formations in question. However, as I establish in this chapter, there is, in fact, only one of the three traps that makes a successful contribution to entrapping "the bad guys" in the case of Loki, namely what I call the "identity/time trap". The mission formations' two other traps "flip over", it seems, entrapping "the trappers" and their desire for action. The result, as I show, is military inaction and existential impotence, challenging cohesion and effectiveness.

ISAF mission formations: Ethnographic recordings

The Army's Monument to the Fallen in Helmand: A very brief history of ISAF mission formations

In its heyday, Camp Bastion hosted a number of coalition bases "inside the wire". Among these was Camp Viking – Denmark's main logistics hub in Afghanistan. At the heart of the base, across the parade ground from the main gate, you would find a war memorial: a stone cairn crowned with a brass cross. Toward the end of ISAF's mission, you would find 38 names engraved in small brass plates, all mounted on the cairn. You would find the 38 names of those Danish soldiers who had lost their lives in the Danish Helmand campaign; 38 names that bore witness to Denmark's military engagement in the Afghan province from 2006 to 2014. The first name dates back to 2007, the last to 2013. Denmark, however, lost its first soldiers in Afghanistan back in 2003, and the total number of fatalities among Danish troops in the country amounts to 44.

Denmark joined the US-led "war against terror" as early as in 2002. In January that year, Danish special operation forces deployed to the US-led Operation Enduring Freedom – Afghanistan, and, in October the same year, six fighter jets were deployed as the first of several Danish Air Force contingents. As for ISAF, the first of nine Danish Army rotation teams joined the Kabul Multinational Brigade in February 2002. The coalition force had been formed two months earlier, in December 2001, following 9/11 and the subsequent US-led invasion of Afghanistan in October the same year. The multinational ISAF force was set up in accordance with the Bonn Agreement, adopted by Afghan delegates under UN auspices in December 2001. The aim of the UN-mandated ISAF force was to assist the Afghan (transitional) government in providing security in the new Afghanistan.

Initially, ISAF's mandate was limited to the Kabul area. Yet, in October 2003, the UN extended ISAF's mandate so that it would cover the entire country by the second half of 2006. Consequently, ISAF grew into one of the largest coalitions in history. At its peak in 2011–2012, the NATO-led force was more than 130,000 strong, and as many as 51 nations have contributed to the coalition over the years (NATO, 2015). In the case of Denmark, politicians acknowledged that the situation in southern Afghanistan was more dangerous than in the north and west of the country. Still, the Danish parliament agreed to support ISAF's expansion in the south and decided to increase the number of Danish troops and deploy them to Helmand under British command. Accordingly, the first of 17 army rotation teams, forming DANCON/ISAF in Helmand, went operational in Camp Bastion in July 2006. At its peak in 2009–2010, the Danish ISAF force was more than 700 strong. In August 2014, the last Danish ISAF soldier left Camp Bastion and Helmand altogether.

As ISAF's mission was drawing to a close, you would find endless rows of tan-colored military vehicles throughout the Bastion-Leatherneck base complex, waiting to be flown out of the mission area. In Camp Viking, next to the main gate, you would for a period of time find a number of Danish Combat Vehicle 9035s lined up for "redeployment". The homebound vehicles had been operated by alternating armored infantry companies since February 2010. However, at "war's end", no Danish infantry was any longer deployed to Helmand. As such, the vehicles gave testimony not only to the changing composition of the Danish ISAF force but also to the changing character of ISAF's mission (formations).

The emphasis of ISAF's mission had shifted over the years from combat to capacity-building to redeployment, that is, from engaging so-called "insurgents" to training, advising and assisting Afghan security forces, to withdrawing and disbanding the coalition forces. This development was obviously reflected in the changing numbers and combinations of the types of coalition forces constituting

ISAF's mission formations on the tactical level. In that respect, DANCON/ISAF was no exception. Combat and combat support units conducting counterinsurgency warfare against Taliban and other insurgents on "the front line" were the pivot of the Danish mission formation, from Team 1 to Team 12. Indeed, from Team 4 (October 2007) to Team 12 (January 2012), DANCON/ISAF formed an independent battlegroup that was structured around a combat battalion centered on two combat companies, each about 130 strong. By comparison, from 2006 to 2012, the Danish Civil-Military Cooperation (CIMIC) Detachment was staffed with no more than ten officers and soldiers.

From February 2012, the focus was displaced from combat to capacity-building: the Danish ISAF force no longer made up a battlegroup with its own area of responsibility, and about half of DANCON/ISAF was now tasked with training Afghan security forces. In August the same year, one of the two Danish combat companies was replaced with three police operational mentoring and liaison teams, each staffed with military personnel and civilian police officers. In July 2013, the last Danish ISAF combat company departed from Helmand. From August the same year, the largest unit in the Danish mission formation became an HQ and logistics company, embodying the shift from capacity-building to redeployment. Combat and combat support troops, however, remained an integral part of the Danish ISAF force until the end of the mission.

HQ DANCON/ISAF, RC(SW):
Command structures of ISAF mission formations

In Camp Viking, you would have found the HQ of the Danish ISAF force in Helmand to the right of the Danish Army's Monument to the Fallen in Helmand. On the door into the large tent accommodating the Danish staff, you would have found a sign announcing "HQ DANCON/ISAF, RC(SW)". Thus, the sign might have reminded you of the fact that the Danish mission formation was entangled in a larger command structure, namely, Regional Command South West – RC(SW).

In step with ISAF's expansion and the surge of coalition forces, Afghanistan was divided into altogether six regional commands (RCs). These multinational, military formations, all subordinated to ISAF's commander-in-chief and HQ in Kabul, were of different strengths, ranging roughly from brigade to division size. Each command was in charge of security and reconstruction in its designated area of responsibility. Each command worked in partnership with the Afghan government and the Afghan security forces. Each command had a number of military task forces assigned, just as there were a number of provincial reconstruction teams (PRTs) located within each command. The task forces conducted military security operations in their assigned areas of responsibility, and the PRTs, staffed with civilian and military personnel, worked to support reconstruction efforts throughout numerous provinces across Afghanistan.

From 2006 to 2010, the Danish ISAF force formed part of RC(S), headquartered at Kandahar Airfield. The command's territory initially covered southern Afghanistan's six provinces. However, when the command's strength reached

50,000 coalition forces in the course of 2009, it was split in two to improve its ability to command and control. In July 2010, RC(SW) was stood up. It was headquartered in Camp Leatherneck and assumed responsibility for the security and reconstruction in Afghanistan's two furthest south-west provinces: Nimruz and Helmand. The new command was comprised of contributions from Bahrain, Bosnia and Herzegovina, Georgia, Jordan and Tonga as well as from four NATO member states: Denmark, Estonia, the UK and the US. The command was the home of the Helmand PRT and it had three major task forces assigned: Task Force Helmand, Task Force Leatherneck and Task Force Belleau Wood.

The British-led Task Force Helmand was stood up in April 2006 and was, until August 2013, headquartered in Lashkar Gah, Helmand's provincial capital. Originally, the multinational task force, roughly of brigade size, was charged with a counterinsurgent (COIN) mission. The plan was to pursue an "ink-spot" strategy involving a clear, hold and build approach that would cover the triangle in central Helmand between Camp Bastion in the north-west, Gereshk, Helmand's commercial center, in the east, and Lashkar Gah in the south. The plan was to clear and hold this triangle, or "ink-spot", as a secure zone, allowing nation building to take place through reconstruction efforts (Jakobsen & Thruelsen, 2011). The next step would then have been to gradually expand the ink-spot north of Gereshk toward Sangin in the Upper Gereshk Valley and further north to Kajaki (ibid.), the location of the strategically important Kajaki Dam. However, owing to political pressure from Helmand's governor, Mohammed Doud, and from Afghanistan's President Karzai, the coalition forces soon found themselves spread across roughly 1,550 of Helmand's more than 58,500 square kilometers, from Musa Qala in the north to Garmsir in the south (ibid; King, 2010; Pedersen, 2017b).

By October 2009, the US Marine Corp's Task Force Leatherneck had arrived in theater and assumed responsibility for "the entire western and southern sections of the province, while the remaining part of the province was divided among UK Battlegroups North-West (Musa Qala), North (Sangin) and Centre South (Lashkar Gah) ... and the Danish Battle Group Centre (Gereshk)" (King, 2010, p. 315). All the same, in spite of the surge of coalition forces, the Helmand campaign remained, as King (2010, p. 313) has put it, "defined by the dispersal of forces into isolated forward operating bases (FOBs) and recurrent but indecisive offensive operations against the Taliban".

As for the all-American Task Force Leatherneck, it formed part of the US military surge in Helmand and became the ground combat element of RC(SW). The task force was initially headed by the 2nd Marine Expeditionary Brigade (from May 2009 to April 2010) and then alternately by the 1st and 2nd Marine Divisions (seemingly from April 2010 to March 2013). On its part, the multinational Task Force Belleau Wood was formed in early 2011. As with the case of Task Force Leatherneck, the brigade-sized Task Force Belleau Wood was headquartered in Camp Leatherneck and was led in turn by the 1st and 2nd Marine Expeditionary Forces until the end of the mission in October 2014. The task force was charged with force protection, that is, with defending the perimeter of the Bastion-Leatherneck base complex. However, the mission of the task force was not simply to prevent attacks on the two camps by maintaining a constant presence in the battlespace in terms of standing post, watching for suspicious activity and turning out as a quick reaction force in the face of enemy threats (Solano, 2011). Rather, the task force was assigned an area of operations, spanning more than 1,000 square kilometers, and made an effort to prevent attacks by way of "taking an offensive and aggressive approach toward the enemy, engaging them away from the base out in the area of operations" (Ostroska, 2014.).

A horned helmet: Loki and the tank detachment

When you entered Camp Viking through its main gate, the first work area that you would have come across, on the right, would be the one belonging to the Tank Platoon Loki. At least that would have been the case during ISAF's redeployment phase. The work area was built up around a handful of freight containers with supplies, equipment and maintenance tools. On one or two of the containers, you might have noticed a black spray-painted stencil: a horned helmet. You might have seen the same helmet printed on the tank platoon's obligatory tour T-shirt. You might also have noticed it on the turret of Loki's Cougar MRAP. You might even have spotted it on the platoon commander's tank helmet. You might indeed have recognized it from Marvel's Hollywood production *Thor: The Dark World* (2013), that is, you might have recognized the horned helmet as a devilish-looking depiction of Loki's helmet. As such, the horned helmet was a symbolic reference to the tank platoon's name, its call sign – one that had traditionally been used by Danish tank platoons deployed to Helmand on previous rotations. The call sign "Loki" was taken from the trickster god, the god of evil and mischief in Nordic mythology. In that way, the tank platoon staged itself as present-day Viking warriors and indicated the unit's destructive potential.

At the center of Loki's work area, in the shade of a stretched-out camouflage net, you would find a long wooden table facing a whiteboard. When the tankers were "inside the wire", they would gather around this table, be it for daily updates, maintenance, operational commands, preparations for combat or after-action reviews. Next to the freight containers encircling the long table, you would find Loki's Cougar MRAP parked to one side and its four main battle tanks, its four Leopard 2A5s, on the other. "Outside the wire", Loki was deployed with the Cougar and three of the Leopards. Each vehicle was manned with four crew members: a vehicle commander, a driver, a gunner and, in the case of the Leopards, a loader, or, in the case of the Cougar, a signalman. Each Leopard was armed with a 120-mm tank gun and a 7.62-mm light machinegun. Obviously, it was the former that secured Loki the position as the unit with the hardest-hitting weapon system available among ISAF's ground forces in Helmand. The Cougar, on its part, served as Loki's communications center and was armed with a 12.7-mm heavy machinegun.

Apart from the platoon commander, First Lieutenant (1st Lt) Frederiksen, all vehicle commanders were non-commissioned officers (NCOs). The rest of the tankers were all corporals and privates. All crew members were male, most were in their mid- or late 20s, many were in a relationship, and some had children. Most of the tankers were unskilled workers before they joined the Danish Defence, a few were fully qualified workmen, and some had volunteered upon completing their upper secondary education. As Loki formed part of the Danish Army's standing force, all crew members were regulars, and all, except for the first lieutenant, had previous deployment experience. However, it was only about half of the crew who had previously been deployed with a tank platoon. Thus, although the desire for action broadly speaking was not as outspoken within the ranks of Loki as it was in the case of Fenrir's younger, first-time-deployed personnel, it was not exactly conspicuous by its absence. After all, some were back for more, whereas others were longing for their first time in action with a 62-tonne main battle tank.

If you were to continue past Loki's work area, be that to the left or right of the main gate, you would have found the unit-specific work areas of Loki's "enablers" and "assets" spread out along the perimeter of Camp Viking. The enablers included mine-clearing, engineers, tank recovery, combat medics and supply transport. The assets consisted of an "ISTAR package", that is, a mission formation combining intelligence, surveillance, target acquisition and reconnaissance tasks with the aim of improving situational awareness and the basis for decision-making. Loki's ISTAR force was a multi-element force composed of an electronic warfare (EW) team, an unmanned aircraft systems team and the TACP mentioned in the introduction. Operation-specific add-on assets included military police, a liaison officer and an intelligence officer.

Along with the tank platoon, the enablers and assets made up the constitutive elements of a combined force, namely DANCON/ISAF's tank detachment. The detachment was headed by Loki, was usually 50 to 60 strong and consisted of 12–14 vehicles that formed a convoy stretching out for some 1.5 km during tactical driving. Neither the tank platoon nor the tank detachment operated on their own in theater. Loki and its enablers and assets were all inseparable elements, constituting the tank detachment as what I call a "plug-and-play unit", that is, a NATO-compatible force available for secondment to fellow NATO and coalition partners as an indivisible unit, however composite it may be. By implication, DANCON/ISAF can, with Ben-Ari, Ben-Shalom, Brønd and Padan (this volume), be compared to a holding company among whose constituent elements the tank detachment was detached from the Danish contingent just to be attached to other coalition forces, thereby forming new mission formations.[6]

In total, the Loki-led tank detachment was deployed to 16 operations, amounting to 68 days and nights away from Camp Bastion. A typical operation would run for three or four days. Thirteen operations were carried out under the command of the UK-led Task Force Helmand, and three were conducted under the US-led Task Force Belleau Wood. The first half of Loki's tour of duty overlapped with the British Army's deployment of the 1st Mechanised Brigade to Operation Herrick 18.[7] During this period, the tank detachment was attached to the Household Cavalry Regiment's ISTAR Battlegroup that formed part of Task

Force Helmand's Brigade Reconnaissance Force. In addition, the tank detachment was on one occasion deployed to an operation under the command of the First Fusiliers Battle Group – stood up by the Royal Fusiliers Regiment's 1st Armoured Infantry Battalion. As for the second half of Loki's deployment, it coincided with Operation Herrick 19. The 1st Mechanized Brigade was replaced with the 7th Armoured Brigade, also known as the "Desert Rats". The Fusiliers were succeeded by a mechanized infantry battalion from the Royal Regiment of Scotland, the 4th Scots, also known as the "Highlanders". Loki was, along with the rest of the Danish tank detachment, joined to the Highlanders' Manoeuvre Battle Group. On top of that, the tank detachment joined three Task Force Belleau Wood operations with the 9th Marine Regiment's 1st Infantry Battalion (1/9 Marines), also known as "The Walking Dead".

Entrapments in Danish ISAF operations: An anthropological analysis

Trap 1: Doctrines/commanders

BRAC-T, First Fusiliers Battle Group, FOB Ouellette, Task Force Helmand

Loki's last operation under the command of the First Fusiliers Battle Group was an 8-day tour to the British FOB Ouellette, about 60 km east of Camp Bastion, across the Helmand River. It was a "base relocation and closure – transition" (BRAC-T) operation[8] in which the Danish tank detachment overwatched the surrounding terrain, while Outpost Dara was demolished and FOB Ouellette subsequently vacated and transferred to the Afghan National Civil Order Police (ANCOP). After the operation, I interviewed 1st Lt Frederiksen back in Camp Viking. He had grown a mustache since we met last. It roused my curiosity. What was the story behind it? Was there a connection to the Ouellette operation? This is what Frederiksen told me:

> He [the battlegroup commander in charge] wouldn't let us go. We couldn't be allowed to get away from up there. Never in my life have I met people that were that bad at planning things. For a start, we had said … we leave from there at nine o'clock no matter whether these ANCOP people have arrived or not … That maneuver … that's what's called "takeover in place" or "relief in place" … There's a standard procedure for that. There's even a NATO paper describing what one has to do, and I do know the British have signed it … Nonetheless, they've decided to throw it to the winds because there were very basic things that were not coordinated … Nine o'clock, still nothing. Then they postponed until half past noon … The clock turns 1 pm, 2 pm and 3 pm as well, and at one point Søren [Sgt Eriksen, Loki's Cougar commander] calls in on the radio – we were pretty frustrated – … "six zero, I just have to ask, those mighty big balls that ISAF had earlier on today when we said not a second later than 1 pm have they just been kicked up in our throat by some

date farmer from down here?" There was no reply on the radio, and then we knew what had happened ... I've then grown a snail as a compliment to him [the commander] ... When I'm positive that he's on a plane back to England, I'll remove my mustache. I've no problem about going around looking like a fool in his honor for the next many weeks.

Frederiksen and the tank detachment, we can say, became entrapped in the mission formation of the Ouellette operation, in its field of tension between doctrinal "textbook war" and the actual "British way of war", as it was waged under the First Fusilier Battle Group commander. Loki was kept waiting for hours, robbing the crews of their sense of agency and striking a blow to their quest for existential potency. What is more, Loki was not only left waiting by the outbound British commander, but also by his inbound Afghan relief. The latter, to judge from Sgt Eriksen's radio message, challenged the moral order that should ideally exist between senior and junior persons, between rulers and ruled (Jackson, 2005), between "mentors" and "mentees" and between ISAF and Afghan National Security Forces. Along this line of thinking, we can see Frederiksen's mustache as a way of symbolically turning the tables, a way of re-establishing the moral order and reclaiming a sense of existential potency.

Command, Manoeuvre Battle Group, Camp Bastion, Task Force Helmand

Operation Herrick 18 turned into Operation Herrick 19. The Highlanders arrived in theater, the Fusiliers redeployed, 1st Lt Frederiksen shaved off his mustache, and the tank detachment became attached to a new mission formation: the Manoeuvre Battle Group. The new battlegroup commander, Lieutenant Colonel (Lt Col) Roddis, and his operations staff continued where their predecessors had let off. In fact, from Loki's perspective, things, it seemed, went from bad to worse. After all, it was not without reason that the lieutenant colonel, within his own ranks, had earned the nickname "Darth Roddis and the Manoeuvre Death Star":[9] reportedly it matched his style of command that appeared more akin to the individualistic, even heroic, approach of the last century than to the collaborative efforts defining command in the 21st century (King, 2019). Naturally, the Danish tank crews were not slow to pick up the nickname; meanwhile, they became further entrapped in the mission formation's discrepancy gap between NATO doctrine and the "British way of war", let alone in the mission formation's multinational composition and its resulting hierarchy, with a high-handed British commander at the top. Obviously, such entrapments gave rise to a fair share of puzzlement, shaking heads and sarcasm within the ranks of Loki. In the words of 1st Lt Frederiksen:

Now, we've counted five operations under this Manoeuvre Death Star, and little by little a picture has begun to emerge ... Now, when one is in the process of redeploying, and the English no longer are that happy about getting their people shot and burned to pieces, then they've found a brilliant solution, right? They can take an Estonian infantry company,[10] and send it to its death, and then they can take a Danish tank platoon to stand and watch in the meantime. All of it then sail under the flag, the proud Union Jack … I get the impression that the only reason I'm the second maneuver element in a battlegroup is because … then he [the battlegroup commander] has two maneuver elements on the ground, and then he's obliged to decide on the coordination of fire and movement that has to take place between myself and the Estonians … That's what a commander is doing.

However, much to the frustration of the tank commander, Warrant Officer Class Two (WO2) Overgaard, the battlegroup commander did not limit himself to resuming control of fire and movement between the maneuver elements:

When you then take into account that Roddis, the commander of the Manoeuvre Death Star, he feels that he's the one who has to take charge when, for instance, a mine plough has been blown up … It's then I think to myself, "that can bloody well not be true". How can they [the British] ever have been playing a part in defeating the Germans? How the hell can they ever have had the entire India and great parts of Africa under their control?

1st Lt Frederiksen picked up on the imperial legacy and linked it to what he regarded as leadership arrogance in the battlegroup and to what he perceived as a stubborn aversion to complying with NATO doctrine:

Once they [the British] ruled more than one-third of the known world … It's as if they're used to waving the hand a bit, and then the hordes will be sure to fall into line … That Manoeuvre Battle Group … It had that arrogant British attitude that now we should move out into to this area here, and then they [the British] should just be making the entire plan … The Estonians did actually not do what the British wanted them to do. However, they did what made sense because they know the area, and … because the British ignored all input the Estonians put forward … So, the Estonians did as they saw fit, and that was a great thing to see … They were damn keen those boys. They were competent people … We [the Danish army] have educated the Estonians back in the 1990s … So, we do just totally agree on how to do things … When I receive an English plan, then there's one or another obscure effect sketched in [such as, deter or understand] … Karl [the Estonian company commander] and I agree that's not the right way to do it … But this stubborn persistence on that's the way it's going to be because they're the head of the snake – it implies that we must adapt. It just means that there's a greater risk of misunderstandings … The British think they comply with mission command, *auftragstaktik*, which is what we use ourselves, and which everyone in NATO is using. There you say you want a task done … and there follows some desired effect … the British just give the effect without assigning a task … It just doesn't make any sense.

Taken together, the four interview excerpts in the present section form a picture of what I term the "doctrines/commanders trap". It is a dual trap "set up" by battlegroup mission formations and designed for entrapping "the enemy" on the basis of shared doctrinal procedures and command capabilities. However, as the excerpts show, the multinational dimension of the mission formations, so to speak, makes the trap "flip over" and it entraps Loki in different doctrinal/command practices, challenging the effectiveness and cohesion of the tank detachment as well as the battlegroup as a whole. Furthermore, as indicated in the case of the Ouellette operation above, the "doctrines/commanders trap" reduces the sense of existential potency among the entrapped tankers. The same holds true for the command of the Manoeuvre Battle Group. First, if we are to believe 1st Lt Frederiksen's account, then Loki was deployed as a passive witness to the fate of the Estonian infantry on the ground as if the presence of the Leopards did not count. Indeed, Frederiksen suspects that Loki was only deployed to legitimize the execution and command of operations within the battlegroup framework. Second, if we follow WO2 Overgaard's story about the micro-managing battlegroup commander, then the commander was virtually declaring Overgaard and his fellow countrymen incapable of managing what Overgaard regarded as rightly should have been their own affairs rather than those of a lieutenant colonel. Third, according to 1st Lt Frederiksen's description of the British monopolization of decision-making, the Estonians and the Danes were not allowed to have a say in the operations planning. Moreover, on the basis of Frederiksen's detailing of the British variation of mission command, we can say that putting forward a desired effect of an operation, rather than clearly defined operational tasks to be carried out, harbored the risk of making the Estonians and the Danes incapable of action or, at least, incapable of taking the right action.

Trap 2: Rules/systems

Rules of Engagement, HQ ISAF, Kabul

There was some 700 km between Camp Bastion and HQ ISAF in Kabul. Yet, the presence of ISAF's commander-in-chief (COMISAF) was in one sense all-pervading at the tactical level of the ISAF mission formation across Afghanistan. In an effort to minimize "collateral damage", successive commanders-in-chief had, since 2009, recurrently tightened ISAF's mission-specific ROE (Beljan, 2013). Indeed, in 2009 and until the end of the mission in 2014, COMISAF had made it mandatory for all ISAF troops to regularly participate in "judgmental training" in the legal use of military violence (Pedersen, 2017b, 2019b). As such, COMISAF's approach to civilian casualties at the hands of the coalition forces was embodying the growing juridification of the "Western way of war" (Jones, 2016; Pedersen, 2017b, 2019b; see also Ben-Ari et al., this volume). "Inside the wire" this meant that Loki, every fortnight or so, found itself in judgmental training classes instructed by the Danish military legal advisor. "Outside the wire", self-defense aside, Loki's resort to offensive use of force was restricted to NATO ROE 421–424, that is, to engaging "targets" who either performed "hostile acts" or demonstrated "hostile intent". Crucially, ISAF's strict ROE did also entail that the application of 60-mm weapon systems and larger had to be authorized at the level of brigade commander. Thus, Loki had to transmit a request up through the chain of command, asking for permission to open fire with the 120-mm tank guns. This condition, along with the other violence-restricting rules, was not exactly appreciated within the ranks of the tank platoon. As 1st Lt Frederiksen put it in his distinctive sarcastic way:

> There're no reasons why those people ["Taliban" militants] should be allowed to live. They're just bad guys. They're disgusting, vicious, mean people … Because we're obliged by our own ridiculous morality, we cannot just simply move out and cut the throats of these people … We'll have to do some paper-work, or else we will end up in the Hague in a few years.

Specifically addressing the restrained use of Loki's primary weapon system, Frederiksen stated:

> That chain of command, it's so long and difficult … There's nothing in the world that I can say on the radio that can ensure a brigade commander that that one there, that's a bad guy … So, you can have 600 men on the ground, but if it's still the general, one single person, who has the authority to decide whether we may open fire or not, then we're just magnets for enemy fire … Then we can just sit and wait for them [the insurgents] to encircle us and make an ambush, or what do I know.

On the basis of these two interview snippets, we can tell the rough outline of what I have coined the "rules/systems trap". In the present context, it is a two-sided trap tailor-made for entrapping insurgents in accordance with ISAF's ROE and with the application of Loki's weapon systems. Yet, as with the case of the doc-trines/commanders trap, the rules/systems trap tends to "flip over", entrapping the tankers in the juridification of war as well as in their own "technological fix" to a "war amongst the people" (Smith, 2005); a war ideally conducted as "precision warfare" (Ben-Ari et al., 2010) to honor the demands for minimizing the numbers of civilian casualties. Accordingly, to judge from Frederiksen's accounts above, the tank crews found themselves vulnerable to "enemies" closing in on them as well as existentially impotent in the face of "the evil" that insurgents brought into the world.

Find and understand, Manoeuvre Battle Group, Task Force Helmand

Loki's crews did not only become entrapped in ISAF's ROE and in the destruc-tive potential of their own primary weapon system. Occasionally, the tankers were also entrapped in the very mission formation of the tank detachment owing to what we can call the "rule of indivisibility". In theater, radio traffic made up a source of intelligence that the tank detachment's EW asset tapped into in order to monitor radio communication among the insurgents within range. However, the EW team did not only provide information about imminent dangers to the troops on the ground but also to the intelligence gathering in "find and understand" operations.[11] Especially, the latter input apparently made the EW team a highly attractive asset for the Highlanders' Manoeuvre Battle Group. Yet, as the tank detachment formed an indivisible "plug-and-play" unit, the EW team could not be separately detached. Hence, the complete tank detachment had to be deployed, even though the British, it seems, did not take much interest in the rest of the Danish ISTAR package, let alone in the rest of the Danish tank detachment in "find and understand" operations. This favoring of an asset over Loki was not well received among the tankers. In the words of 1st Lt Frederiksen:

> It's not like that we by all means have to go out and shoot at some people, but sometimes we've had the feeling that they [the British] deploy the tank platoon because, in the detachment, there're these bloody EW people who can listen in on Taliban's radios ... If the three tanks and the MRAP are deployed, then we must take the engineers along, the combat medics, the mine-clearing and the recovery vehicle ... It's absurd that we must "risk" 40–45 men in order for six men can sit out there and do something with a radio. I don't think that's worth it ... It's unfair.

WO2 Overgaard was not a great fan of "find and understand" operations either:

> I've rarely been taking part in anything that foolish. For a start, we do still drive out without being allowed to shoot ... Sometimes we even just drive out in order to listen with this EW as main effort. When we're then in position and ready, then the English come driving by with their ECM [electronic countermeasures] turned on with the result that we cannot hear anything anyhow. Then one feels a bit like one's jaw is dropping.

As the above two interview excerpts indicate, both Frederiksen and Overgaard felt somewhat double-crossed owing to the British deployment of Loki to "find and understand" operations, where the role of the tank platoon was reduced to a walk-on part, striking yet another blow to the tankers' quest for existential potency. Indeed, as Frederiksen's objection points out, the tank crews still ran the risk of having their very being reduced to nothingness, in spite of their subordinate and, therefore, less heroic role. Moreover, Overgaard seems to have felt double-double-crossed when he experienced the British ECM blocking the radio traffic, thus undermining the very purpose of the operation. In a nutshell, the UK-led "find and understand" operations caused division within the Danish tank detachment. Pushing it to the extremes, we can say that the EW asset turned out to be anything but an asset for Loki, thereby putting a strain on the cohesion and effectiveness of the detachment and, by extension, the battlegroup.

Trap 3: Identity/time

COIN patrol, Operation Grizzly XIII, Ismat Bazaar,
Task Force Belleau Wood

The Danish tank detachment was deployed with Task Force Belleau Wood's 1/9 Marines in three operations, all COIN patrols,[12] and all under the command of the Chronos Company. Never before had Loki operated side by side with infantry from the United States Marine Corps (USMC). However, the tankers had great expectations, not least owing to Hollywood productions and military history books. Also, the call sign of the 1/9 Marines, the "Walking Dead", did not exactly lower those expectations. Likewise, the call signs of the battalion's three companies were not Alpha but "Anger", not Bravo but "Bad", not Charlie but "Chronos".

All the same, the two first operations largely failed to fulfill expectations. The tankers were not impressed. In fact, the first operation, called "Grizzly", convinced the tankers that it was not exactly the A-team the marines had fielded. First, there was the incident with the lost marines. One night, three marines turned up at the tank detachment's makeshift camp, its "harbor".[13] They were looking for their platoon. They were lost, and a tanker on guard had to point them in the right direction. At least, that is how the story goes: "It was like a scene taken out of a Second World War movie", 1st Lt Frederiksen subsequently told me, laughter in his voice. Second, and more disturbingly, there was the "blue on blue" incident. Loki was providing overwatch to Chronos infantry on the ground. Small arms fire. Troops in contact. Chronos engaged "the enemy". Yet, from Loki's position on the high ground, it soon became clear that it was a case of friendly fire. Two Chronos platoons were exchanging fire with one another. What was more, one of the platoons was marking "the target" with smoke and requesting Loki to open fire. After that incident, the Chronos company commander was replaced, and the unit had to undergo further training back at Camp Leatherneck before it was allowed "outside the wire" again. "They were simply not good enough", Frederiksen concluded.

However, before the commander of Chronos Company was discharged, he actually managed to enter Jutland Dragoon lore. On the second day of Operation Grizzly, a Chronos platoon commander was overlooking the tank detachment's harbor and spoke about it on the company radio net as the "gypsy camp". Bad idea. The company commander broke in on the radio and tore his platoon commander apart for everyone to hear. Basically, he was told that "the Danes were greater warriors than he would ever be", Frederiksen recalled. Next day, when the vilified platoon commander reported on the company net that his unit was running out of fuel, Sgt Eriksen, Loki's Cougar commander, could not help himself from rubbing salt into the platoon commander's wounds: Eriksen called in on the radio net, kindly offering the commander his assistance, yet Eriksen did so without using his proper call sign. Instead, he identified himself on the net as the "gypsy king".

What are we to make of such incidents and the ways they are recounted? They are not simply anecdotes. They are stories based on the transient cooperation implicated in today's multinational mission formations. They are stories dealing with the discrepancy gap between the expected and the experienced, between mytho-histories and first impressions. They are stories by which warriorhood is performed by virtue of positioning self and others as more or less professional in military terms and as more or less potent in existential terms. They are stories by which the Chronos Company, in the case of the lost marines, is rendered as less competent than the tanker guiding them on their way. They are stories by which the Chronos Company, in the case of the blue on blue incident, is conveyed as reckless, even trigger-happy, and thereby as rather unprofessional, even dangerously close to rendering themselves existentially impotent with the request for tank fire. They are stories by which the Chronos Company, in the case of the platoon commander making the gypsy camp remark, is at first portrayed as regarding himself as more potent and professional than the tank detachment, yet in the end is made into an impotent fool, first when the company commander "set things straight" with his "warriorization" of the Danes and then second, when Sgt Eriksen turned the tables through his appropriation of the "gypsy identity". In short, they are stories by which the storytellers reclaim a sense of agency (Jackson, 2013), a sense of existential potency in the world.

COIN patrol, Operation Leopard Shield 2, Patrol Base Boldak,
Task Force Belleau Wood

Loki's third and last operation with the 1/9 Marines was also the tank platoon's final operation during its 6-month tour of duty. The 4-day operation, named "Leopard Shield", saw Loki deployed, along with a Chronos platoon, in the vicinity of the USMC's Patrol Base Boldak, about 10 km south-east of the Bastion-Leatherneck base complex. The aim of the operation was to find, engage and clear the area of insurgents who had directed an increasing number of attacks against the marines (Antonsen, 2016), practically besieging Patrol Base Boldak for some time. On the first day of the operation, the Chronos platoon had troops in contact, and, in a separate attack, a quick reaction force suffered two casualties on its way to reinforce the engaged platoon. The next two days passed without any incident. However, on the fourth day of the operation, the Chronos platoon came under sniper fire. This time, Loki was able to move into a position from where two Leopards could just barely get "the target", a 2-strong sniper team riding a motorcycle within range. After all, the insurgents drove in and out of sight between compounds, apparently in search of a new position to reengage the Chronos infantry. A 120-mm high-explosive anti-tank round was released from each of the two Leopards in position, defeating "the target" from a distance of 2,130 meters.

This engagement, I contend, was intimately tied to Operation Grizzly. Why? Because the gypsy camp incident displaced the expectation pressure from Chronos to Loki. To be praised as great warriors is to be challenged to prove those words to be true. They will be empty words if one does not stage oneself as warrior, put action behind the words and prove oneself to others by demonstrating one's existential potency within the bounds of one's military professionalism. In this light, we can understand why Loki engaged the sniper team, even though the two insurgents neither had made it to a new position in the moment of the engagement nor made up a "target" that was easy to hit without the risk of causing "collateral damage". Yet, in a different light, we can see that Loki's defeat of two insurgents might not only have to do with competition for warriorhood, provoked by the mission formation's multinational character. Rather, it might also have to do with the very same mission formation's temporariness. After all, the engagement was the first and the last during Loki's deployment and took place not only on the last day of Loki's last operation but also on the very last day in the very last hour that the tank platoon was serving as an operational unit in theater. As Sgt Eriksen phrased it:

> It was a release for the entire rotation team. The main battle tanks have not opened fire in two years,[14] and then we finish right there in the eleventh hour ... So, in part it's fab because it's in the very last hour ... And then in part it's fab to complete the mission, and the relief that now we've done what we have been down here to do. That's awesome. There were cheers. We were giving high-fives and shouting in the car. It was just fab.

Tank gunner Lance Corporal (LCpl) Nielsen elaborated:

> They had been asking for trouble out there in Boldak. The Americans had taken a bit of a beating for some time ... So, the mood that had been built up was a bit like this one [this operation] should bring about combat ... I recall that we shouted with joy just when we had released fire ... We had taken some of these evil people out. On this tour in particular, we have seen so many evil people, but we've not been allowed to shoot anyone. And the frustrations have kind of just grown and grown, and then at last on the same day, which is our last time outside the wire, we are then allowed to shoot someone.

Finally, 1st Lt Frederiksen added:

> We got to shoot two Taliban [fighters]. ... It was a tremendous relief. In part because we had now been down to support someone, and now we had actually been supporting them by killing someone, and they [the Americans] were happy about it because it would give them some peace and quiet in their area for some time ... I was happy to match my responsibility. I've proven that I'm made of the stuff it takes.

From these three interview excerpts, it appears that the killing of the two insurgents made the tankers existentially fulfilled, however temporarily. The engagement, we can say, completed the tankers' quest for existential potency in theater. At least in the case of their last ISAF tour. This sense of completion, of fulfillment, seems to have been augmented, as Sgt Eriksen's description underlines, not only by the fact that the engagement took place in the nick of time and after a weary journey over the past six months. Rather, the sense of existential potency was also strengthened by the fact that Loki's engagement was the first of its kind for some time. For once, as it follows from LCpl Nielsen's account, the tank crews were not reduced to passive witnesses in the presence of insurgents taking advantage of ISAF's strict ROE. For once, the tank crews were allowed to act, for once they were allowed to get into action. For once, as 1st Lt Frederiksen points out, the tankers made their presence count directly and with an effect that would last for some time after the Leopards had left the high ground. Moreover, as Frederiksen appears to emphasize, he felt existentially fulfilled because he had proven to himself and to others that he could assume the responsibility of platoon commander when it came to the crunch. He could perform the warrior identity when it counted. He did not freeze. He did not flee. He gave the order to open fire.

From the accounts of Operation Grizzly and Operation Leopard Shield above, we can now see the initial contour lines of what I have termed the "identity/time trap". It is a trap materialized in the Loki-Chronos mission formation, a trap designed to entrap insurgents through the performance of warrior identities and the race against time among the "trappers". Unlike both the doctrines/commanders trap and the rules/systems trap in the present context, the identity/time trap actually tends to promote military action. It tends to encourage the use of military violence owing to the mission formation's multinational and temporary character: multinationalism stimulates competition between nations to stage oneself as more potent and professional than one's coalition partners, while temporariness prompts one to resort to action while one has the chance in time and space to do so.

Conclusion

In the present chapter, I have anthropologically explored the intimate entanglements of soldierly motivations, military action and mission formations in post-9/11 Afghanistan. Based on ethnographic fieldwork with Danish "grunts" in the temporary, multinational and tailor-made ISAF force, I have endeavored to contribute to the emergent agenda of the new sociology of mission formations and, by implication, to the renewal of the classic sociology of motivation and combat. Empirically, the chapter has been centered on a tactical combat unit, namely the Tank Platoon Loki, and I have put forward an ethnographic account of the changing and partially overlapping mission formations in which Loki took part in theater, ranging from commander-in-chief and regional command over task forces, battlegroups and companies to rotation team and tank detachment. I have demonstrated that the multinational character of Loki's mission formations tends to promote cultural dynamics leading to competition for warriorhood and conflicts over different doctrinal and command practices. I have shown that the juridification and technologization of war at the macro level affect Loki on the ground, be it its individual crew members, its collective mission formations or its social organization of violence. I have indicated that the temporariness of Loki's mission formations encourages military action in a race against the clock, a race against the end of operation, end of tour and end of mission. Methodologically, the chapter is built on my fieldwork as an "embedded anthropologist" at the tactical level in theater, and I have offered a "grunt ethnography" on mission formations in a "view from within", that is, I have presented a study of mission formations in a first-person perspective from within the ranks of Loki's crews. Regrettably, I was denied access to ISAF operations in theater, and I call for further ethnographical research to be conducted, embedded in current and future mission formations during operation planning, as well as during the actual execution of the operations. Theoretically, the chapter has drawn upon existential anthropology and the anthropology of traps. Elsewhere, I have identified the desire for action as the primary motivation factor among the vast majority of my interlocutors in the Danish ISAF force (Pedersen, 2017a, 2017b, 2019a). In the present chapter, I have conceptualized this desire as a struggle for existential potency, for presence and significance, be that in the battlespace or in the world at large. I have conceived mission formations as traps/entrapments, as material configurations of constituent elements making up specific traps and as embodied processes of entrapping specific "kinds of prey" in specific ways. I have argued that the quest for existential potency becomes entangled and entrapped in mission formations charged with the social organization of violence on the tactical level. "The trappers", so to speak, become existentially trapped in their own trapping devices/entrapment dynamics. In that respect, I have identified the rough outlines of three dual traps of contemporary mission formations: doctrines/commanders, rules/systems and identity/time. In the case of Loki, only the latter trap appeared to make a successful contribution to entrapping "the prey". The two other traps, on their part, tended to flip over, entrapping "the trappers" and their desire for action. Accordingly, as I have shown, military inaction caused existential impotence, challenging the mission formations' cohesion and effectiveness.

This was not only so in the case of Loki and its ISAF mission formations. It was definitely also the case with Fenrir (Pedersen, 2017a, 2017b, 2019a). In fact, the unit was on the verge of mutiny after having been almost completely stuck "inside the wire" throughout the first half of its tour of duty. In the end, Fenrir did not come anywhere near any action, and, contrary to Loki, the unit largely left Helmand frustrated and existentially unfulfilled.

Evidently, this chapter merely constitutes the first tentative beginnings of an anthropology of military mission formations, just as it only makes a small step toward an "anthropologization" of motivation and combat. Essentially, the theorization of mission formations as traps/entrapments needs to be further developed, just as the three existential traps I have identified need to be further elaborated. Research also needs to be carried out on additional traps, such as military trophy hunting and military occidentalism. Political ideologies, such as democracy, might also constitute yet another trap to delve further into. After all, as WO2 Overgaard replied without batting an eyelid, when I – toward the end of "ISAF's war" – asked him about the prospects of democracy in Afghanistan: "Well, then one should just go on with shooting them ['Taliban' militants], and keeping them down out here".

Acknowledgments

This chapter is based on empirical data that I have co-produced with Danish officers and soldiers as part of my doctoral studies, forming part of the collaborative research project *Soldier and Society: Anthropological Perspectives*, University of Copenhagen (2012–15). I would like to extend my sincere gratitude to Eyal Ben-Ari, Uzi Ben-Shalom, Thomas Brønd and Carmit Padan for inviting me to think with "mission formations" and contribute to this edited volume. I owe special thanks to Thomas Brønd and to 1st Lt Frederiksen for their generous comments on a previous draft of this chapter.

Funding

The data generation on which this chapter is based was made possible thanks to the support of my doctoral studies by the Independent Research Fund Denmark | Humanities (FKK) (grant no. 0602-02345B).

Notes

1 MRAP is an abbreviation for Mine-Resistant Ambush Protected. In 2006, the US Department of Defense launched the MRAP Vehicle Program in response to the growing threat from improvised explosive devices in the wars in Iraq and Afghanistan (Browne, 2016).
2 In order to safeguard the anonymity of my research subjects, the stated names of my interlocutors are not their real names but pseudonyms. In keeping with a cavalry tradition among hussars and dragoons, privates are named after their hometowns, such as "Nordrup" and "Thorsminde".
3 "Kill TV" refers to gun camera footage showing Apaches killing people with Hellfire missiles or with the 30-mm chain gun.
4 "Grunts" are shorthand for low-ranking, frontline soldiers and officers in the combat arms.
5 In theater, Fenrir was an integral part of a composite support unit: the Danish HQ & Logistics Company formed by 1st Logistics Battalion, the Danish Logistics Regiment. Occasionally, Fenrir was seconded to British-led operations on an ad hoc basis.
6 In addition to the tank detachment, DANCON/ISAF was holding two other detachable elements: the force protection section Fenrir and the explosive ordinance disposal team Brimstone.
7 "Operation Herrick" was the codename for the UK's war in Afghanistan between 2002 and 2014.
8 BRAC-T operations implied the closure of coalition bases, be that by demolition or by transferral to Afghan security forces.
9 The nickname "Darth Roddis" refers to Darth Vader, a ruthless and much-feared imperial ruler in the Star Wars sci-fi universe created by George Lucas. The "Manoeuvre Death Star" is named after the saga's "Death Stars", that is, the imperial space stations and galactic superweapons.
10 The Estonian infantry company was the Scouts Battalion's Charlie Company. The mechanized infantry battalion forms part of the 1st Infantry Brigade and is garrisoned in Tapa.
11 "Find and understand" operations were basically intelligence operations based on the tank detachment's ISTAR assets.

12 COIN patrols were operations involving find, disrupt and clear combined with key leader engagements with a view to improving relations between local Helmandi leaders and coalition forces.
13 The tank detachment's crews used their vehicles to form a protective ring, a "harbor", providing cover from the wind and insurgents.
14 Sgt Eriksen remembers incorrectly. The last time the Danish tank platoon engaged a "target" was in fact within the past year.

References

Antonsen, T. (2016). *Danish Leopards in Helmand: From the crew's perspective*. Camberley: Trackpad Publishing.
Battistelli, F. (1997). Peacekeeping and the postmodern soldier. *Armed Forces and Society*, *23*(3), 467–484.
Beljan, R. (2013, 30 May). Afghanistan: Lessons learned from an ISAF perspective. *Small Wars Journal*. Retrieved from www.smallwarsjournal.com/jrnl/art/afghanistan-lessons-learned-from-an-isaf-perspective
Ben-Ari, E. (2018). From leading combat units to leading combat formations – Modularity, loose systems and temporariness. In L.V.T. Swedish (Ed.), *Uppdrag Militär: Perspektiv på Militärt Yrkeskunnande* (pp. 53–70). Stockholm: Centre for Studies of Armed Forces and Society (CSMS).
Ben-Ari, E., Ben-Shalom, U., Brønd, T., & Padan, C. (2020). Introduction: Mission formations and a new agenda for the study of military units in action. In E. Ben-Ari, U. Ben-Shalom, T. Brønd & C. Padan (Eds.), *Military mission formations: New sociological perspectives*. New York: Routledge.
Ben-Ari, E., Lerer, Z., Ben-Shalom, U., & Vainer, A. (2010). *Rethinking contemporary warfare: A sociological view of the Al-Aqsa Intifada*. Albany, NY: State University of New York Press.
Ben-Shalom, U. (2020). Research methods for the study of combat formations. In E. Ben-Ari, U. Ben-Shalom, T. Brønd & C. Padan (Eds.), *Military mission formations: New sociological perspectives*. New York: Routledge.
Browne, M. (2016, 1 September). MRAP program celebrates 10 years of protecting those who protect us. Marine Corps Systems Command. Retrieved from www.marcorsyscom.marines.mil/News/News-Article-Display/Article/932752/mrap-program-celebrates-10-years-of-protecting-those-who-protect-us/
Brænder, M. (2016). Adrenalin junkies: Why soldiers return from war wanting more. *Armed Forces and Society*, *42*(1), 3–25. doi:10.1177/0095327X15569296
Brænder, M., & Andersen, L.B. (2013). Does deployment to war affect public service motivation? A panel study of soldiers before and after their service in Afghanistan. *Public Administration Review*, *73*(3), 466–477. doi:10.1111/puar.12046
Desjarlais, R., & Throop, C.J. (2011). Phenomenological perspective in anthropology. *Annual Review of Anthropology*, *40*, 87–102. doi:10.1146/annurev-anthro-092010-153345
Harari, Y.N. (2008). *The ultimate experience: Battlefield revelations and the making of modern war culture, 1450–2000*. Basingstoke: Palgrave Macmillan.
Hoffman, F.G. (2007). *Conflict in the 21st century: The rise of hybrid wars*. Potomac Institute for Policy Studies. Retrieved from www.potomacinstitute.org/images/stories/publications/potomac_hybridwar_0108.pdf
Jackson, M.D. (Ed.). (1996). *Things as they are: New directions in phenomenological anthropology*. Bloomington, IN: Indiana University Press.

Jackson, M.D. (1998). *Minima ethnographica: Intersubjectivity and the anthropological project*. Chicago, IL: University of Chicago Press.

Jackson, M.D. (2005). *Existential anthropology: Events, exigencies and effects*. New York: Berghahn.

Jackson, M.D. (2007). *Excursions*. Durham, NC: Duke University Press.

Jackson, M.D. (2013). *The politics of storytelling: Variations on a theme by Hannah Arendt* (2nd ed.). Copenhagen: Museum Tusculanum Press.

Jakobsen, P.V., & Thruelsen, P.D. (2011). Clear, hold, train: Denmark's military operations in Helmand 2006–2010. In N. Hvidt & H. Mouritzen (Eds.), *Danish foreign policy yearbook 2011* (pp. 78–105). Copenhagen: Danish Institute for International Studies (DIIS). Retrieved from www.diis.dk/files/media/publications/import/extra/yb2011-dan ish-foreign-policy-yearbook_full-book_web_0.pdf

Jiménez, A.C., & Nahum-Claudel, C. (2019). The anthropology of traps: Concrete technologies and theoretical interfaces. *Journal of Material Culture, 24*(4), 383–400. doi:10.1177/1359183518820368

Jones, C.A. (2016). Lawfare and the juridification of late modern war. *Progress in Human Geography, 40*(2), 221–239. doi:10.1177/0309132515572270.

Kaldor, M. (2012). *New & old wars: Organised violence in a global era* (3rd ed.). Cambridge: Polity Press.

King, A. (2010). Understanding the Helmand campaign. *International Affairs, 86*(2), 311–332. doi:10.1111/j.1468-2346.2010.00884.x

King, A. (2013). *The combat soldier: Infantry tactics and cohesion in the twentieth and twenty-first centuries*. Oxford: Oxford University Press.

King, A. (2019). *Command: The twenty-first century general*. Cambridge: Cambridge University Press.

Lyk-Jensen, S.V., & Glad, A. (2018). Why do they serve? Changes and differences in motives of Danish soldiers deployed to peace-keeping and peace-enforcing missions. *Defence and Peace Economics, 29*(3), 312–334. doi:10.1080/10242694.2016.1200220

Lyk-Jensen, S.V., & Pedersen, P.J. (2019). *Soldiers on international missions: There and back again*. Bingley, WA: Emerald.

Matthews, D. (2016, 8 December). The apache helicopter: The army's attack monster. *The Telegraph*. Retrieved from www.telegraph.co.uk/business/boeing-uk/power-of-ap ache-attack-helicopter/.

Mattingly, C. (2012). Two virtue ethics and the anthropology of morality. *Anthropological Theory, 12*(2), 161–184. doi:10.1177/1463499612455284

Mattingly, C. (2014). *Moral laboratories: Family peril and the struggle for a good life*. Oakland, CA: University of California Press.

Moskos, C. (2000). Towards a postmodern military? In S.A. Cohen (Ed.), *Democratic societies and their armed forces: Israel in comparative context* (pp. 3–26). Portland, OR: Frank Cass.

Moskos, C.C., & Burk, J. (2019 [1994]). The postmodern military. In J. Burk (Ed.), *The military in new times: Adapting armed forces to a turbulent world* (pp. 141–162). New York:

North Atlantic Treaty Organization (NATO). (2015). ISAF's mission in Afghanistan (2001–2014). Retrieved from www.nato.int/cps/in/natohq/topics_69366.htm.

Ostroska, J. (2014, 24 February). Task Force Belleau Wood welcomes new Commanding Officer. *Defence Visual Information Distribution Service (DVIDS)*. Retrieved from www.dvidshub.net/news/121179/task-force-belleau-wood-welcomes-new-commandin g-officer.

Pedersen, T.R. (2017a). Get real: Chasing Danish warrior dreams in the Afghan "sandbox". *Critical Military Studies*, *3*(1), 7–26. doi:10.1080/23337486.2016.1231996.

Pedersen, T.R. (2017b). *Soldierly becomings: A grunt ethnography of Denmark's new "Warrior Generation"* (Doctoral Dissertation). Copenhagen: Department of Anthropology, University of Copenhagen.

Pedersen, T.R. (2018). Sidste mand lukker og slukker: Rekviem for en krigerdrøm. *Jordens Folk*, *53*(3–4), 24–35.

Pedersen, T.R. (2019a). Ambivalent anticipations: On soldierly becomings in the desert of the real. *The Cambridge Journal of Anthropology*, *37*(1), 77–92. doi:10.3167/cja.2019.370107.

Pedersen, T.R. (2019b). The entangled soldier: On the messiness of war / law / morality. In B.R. Sørensen & E. Ben-Ari (Eds.), *Civil-military entanglements: Anthropological perspectives* (pp. 164–184). New York: Berghahn.

Pedersen, T.R. (2019c). Breaking bad? Down and dirty with military anthropology. *Ethnos: Journal of Anthropology* [Published ahead of print]. doi:10.1080/00141844.2019.1687550.

Ram, K., & Houston, C. (Eds.). (2015). *Phenomenology in anthropology: A sense of perspective*. Bloomington, IN: Indiana University Press.

Smith, R. (2005). *The utility of force: The art of war in the modern world*. London: Penguin Books.

Solano, T. (2011, 23 December). What is Task Force Belleau Wood? *Defence Visual Information Distribution Service (DVIDS)*. Retrieved from www.dvidshub.net/news/81718/task-force-belleau-wood

Taylor, A. (2013). *Thor: The dark world*. Burbank, CA: Marvel Studios.

9 Logics battlefield

IT contracting and military reserves in the Dutch army

Joseph Soeters, Gerold de Gooijer, Paul C. van Fenema and Nuno Oliveira

Introduction

Despite its very specific *raison d'être* – preventing, containing and solving large-scale violent conflicts – the military is increasingly seen as an ordinary organization (Norheim-Martinsen, 2016). This observation implies that the military has to comply with national and international legal requirements as well as manage businesslike logics added to public values. In particular, the strive for more effectiveness and efficiency underscored by new public management has become influential in the armed forces, as in other public-sector organizations (e.g., Denis, Ferlie & van Gestel, 2015). These developments put the careful use of financial, material and human resources in the military at the center of analysis and policy formulation. The military's actions are no longer determined by the ideas of victory and conquest as the legitimization of whatever military action will cost.

The pursuit of the efficient use of financial means has resulted in an increase in contracting with external parties for tasks that had traditionally been core military jobs. Private military companies (PMCs) have seen unprecedented growth over the past decades, particularly in the United States. PMCs are deployed to hardcore military operations, such as, in recent history, in Iraq and Afghanistan. PMCs conduct activities that are relatively close to the military core business, such as escorting high-risk convoys and securing garrisons, prisons and diplomatic units (e.g., Cusumano & Kinsey, 2015). In addition, PMCs are contracted to provide civilian human resources for mundane jobs such as cleaning, laundry and cooking or – at the other end of the spectrum – for very specialized work, such as IT or other high-tech jobs that require capacity and expertise the military lacks to a varied extent (e.g., Kelty & Bierman, 2013).

External parties are useful in managing the military. These parties do not count in the regular military workforce, and hiring contracts can be terminated swiftly once the operations have ended. External hiring, therefore, reduces fixed costs. In fact, "buying" or outsourcing is a substitute for military hierarchy (Van Fenema & Beeres, 2010). The hiring of external parties increases the military's flexibility, as the military organization can extend and decrease its resources in accordance with the expansion and reduction of operations. This practice of externalizing services, hence, is fully in accordance with the rationales of the economics of organizations, as elaborated by the late Nobel Prize winner Oliver Williamson (e.g., Soeters, 2020).

In much the same manner, reserve forces constitute a so-called flexible layer around the core workforce. Reserve forces consist of citizens who usually have been in the military for a certain period of time and who have started a job in civilian life. Such reserves might be deployed to operations overseas in order to sustain the regular forces for a temporary period. This has become normal practice in today's armed forces, particularly in the US and the UK. The use of reserve forces has also increased considerably in the Netherlands. The number of reserve officers who were part of the contribution of the Netherlands to the ISAF mission in Uruzgan, from 2006 until 2010, was striking. These were civilians with specialized expertise in different fields that were useful for the redevelopment of the region in southern Afghanistan, such as agriculture, education, architecture, as well as water and waste management. Reserve officers were also deployed to provide specific technological expertise.

From a practical and conceptual viewpoint, the deployment of reserve officers brought to light new organizational logics in the military in action. These reserve officers were often employed by for-profit firms and private companies and were "shared" with the military for a specified period of time. While being contracted to two employers, they were confronted with both public/military and business logics. These two logics usually do not go together easily (Jacobs, 1992). A triadic, multilevel relationship emerges between an individual and two organizations (Figure 9.1).

Therefore, this chapter aims to unravel the challenges of the co-existence of public and business logics (i.e., organizational *hybridity*) in the military that emerge from new organizing arrangements. Hybrid organizations are "by nature arenas of contradiction" (Pache & Santos, 2013). By deploying "double-employed" reserve officers to operations overseas, organizational *hybridity* has

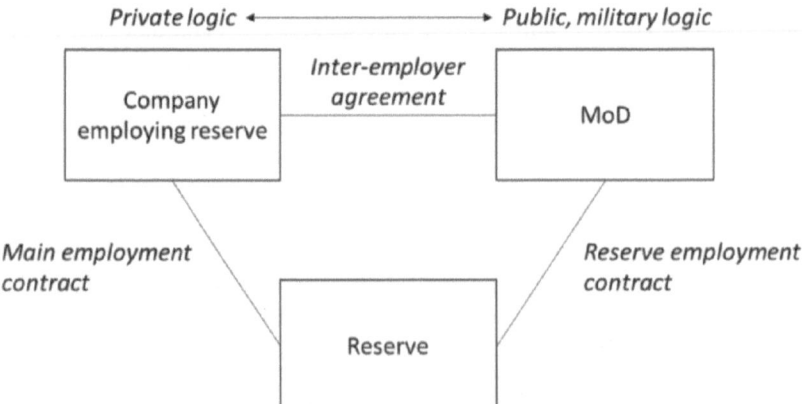

Figure 9.1 Reserves and triadic relationships

entered the military and its operations. Focusing on the IT context (Marschollek & Beck, 2012), this chapter presents a case study of how the Netherlands's armed forces, and in particular its political and strategic leadership, experienced and responded to a crisis as a consequence of this phenomenon. This crisis was dealt with in a manner that solved the crisis, at the expense of ending the advantages of having private-sector expertise inside the military organization in this particular case. The analysis of this case is instructive to derive insights about the challenges of hybridity in the military organization.

Military reserve forces, different organizational logics and hybridity

The increasing reliance on business contractors and reserve forces – the flexible layers of the armed forces – creates its own dynamics and challenges. Contractors and temporary workers are increasingly in demand because their costs are not fixed and can be terminated quickly, and because, in certain domains, there are not enough specialists available. Their hiring is based on an economic logic of cost reduction and follows practices advanced by the business sector, where outsourcing has become a standard procedure (e.g., Bidwell, 2012; Friedman, 2014).

Despite the underlying economic logic, the contract workers' position, in general, is not without rebuttals. Contract workers are appreciated and unappreciated at the same time (Barley & Kunda, 2001). They are appreciated because they not only provide highly needed knowledge and skills, but they also fill in for jobs that are seen as undesirable by the military officials, such as cleaning and sanitary jobs. At the same time, their position in organizations is also relatively weak in a legal and professional-social sense, as they are formally outsiders and are sometimes accused by core personnel of lacking commitment and earning too much money. Their effectiveness depends on rapidly establishing relationships with other personnel and learning about the unique organizational context (Meyerson, Weick & Kramer, 1996). External contractors navigate between receiving respect and experiencing displeasure (Barley & Kunda, 2001, 2006).

Similar observations apply to the position of contract workers in the military. Their increased presence in the armed forces creates tensions and consequences that have not been foreseen. Heinecken (2014) pointed out that the emergence of contractors in the military may mean a loss of the monopoly of knowledge and skills, a loss of autonomy, a loss of the sense of military "corporateness" and erosion of the military service ethic. This may imply "the degradation of the military profession as a whole" (Heinecken, 2014, p. 633). This diagnosis aligns with empirical research in the US military. Kelty and Bierman (2013) studied US core military personnel's perceptions of contractors in Iraq and Afghanistan. They found that the core military displays an attitude of ambivalence toward contractors. They appreciate the value of contractors in leaving them free to engage in combat. Yet they also see contractors making more money yet working less hard, for fewer hours and at a slower pace (Kelty & Bierman, 2013, p. 20). Their presence is also deemed to be detrimental to military customs, traditions and the chain of command. Military commanders lack the authority to intervene when contractors behave in ways that are not conducive to the mission's legitimacy and progress, such as a commander confronted with service-level agreements when ordering immediate water supply (Pijpers, 2013).

Like contractors, reserve forces have a specific "in-between" position in the military. Reserve forces have been associated with "transmigrants", who move back and forth from the private to the public sector of the military (Lomsky-Feder, Gazit & Ben-Ari, 2008). When re-entering the military, reserve forces bring civilian skills and attitudes, as well as experience and innovating capabilities that tend to soften the military (Lomsky-Feder et al., 2008, p. 599). They bring in variety, new ways of thinking, uncommon perspectives. Reserve forces challenge the military world; they "bring about much more critical thinking of 'what is going on' within the military" (Lomsky-Feder et al., 2008, p. 601). Farrell (2010) showed that the substantial improvement in the British actions in the Afghan province of Helmand occurred when a newly formed, but former reserve, brigade entered the arena. This new brigade was less committed to conventional military practices and views. The commanding officers had been traveling between many military and non-military groups and communities throughout their careers. They brought this gamut of experiences into the area of operations and reached results that "traditional" brigades, with all their combat actions, had not been able to obtain. Hence, external contractors can operate in a complementary fashion in military operations.

However, because reserve forces challenge the military world, there are also tensions between them and the regular forces. Reserve forces are criticized because of "skill fade". That is, these forces are "part-time professionals" – that is, a fragmented professional identity – or, in the most derogatory way, they are referred to as the "spare parts" of the organization (Lomsky-Feder, 2008, p. 603). Military personnel are doubtful about whether or not the reserves will turn up and perform adequately when the heat is on. In a similar vein, Kirke (2008) noted that reserve forces in the UK are often seen as "second-class soldiers" who find it difficult to be accepted as operationally trustworthy. This varies, though, as some reserves are as good as regular soldiers, whereas others are deemed to be "useless" (Kirke, 2008, p. 187). In general, reserve forces are more acceptable when their role is farther away from combat, such as conducting maintenance or providing supportive services (e.g., education, sanitation).

Clearly, bringing contract workers and reserve forces into the military implies that different "worlds" enter the organization. Those different worlds may be connected to different *action logics*, or *logiques d'action*, as the French organizational sociologist Lucien Karpik (1972) originally coined the term. Initially, it referred to the different identities, motives and interests that managers and workers have, which may lead to tensions and sometimes even real labor disputes (e.g., Bacharach, Bamberger & Sonnenstuhl, 1996). Nowadays, different action logics usually refer to identity-related behavioral, institutional and cultural patterns in different types of organizations, where the differences between public- and private-business-sector organizations are most striking.

Given the current practice of outsourcing and making use of reserve forces, private-sector action logics – business logics – enter military organizations. This is a fascinating development, as armed forces traditionally adhere to the public, "guardian moral syndrome" as opposed to the "commercial moral syndrome" (Jacobs, 1992, p. 215). Yet, the boundaries between those two "syndromes" tend to become porous nowadays, as both types of organizations approach and sometimes really interpenetrate each other. This undoubtedly creates organizational hybridity, "strange" combinations of action logics inside one and the same organization or organizational configuration (Batillana, Besharov & Mitzinneck, 2017; also: Pache & Santos, 2010, 2013). The question is: how can these two logics go together versus when vulnerabilities surface?

Researchers identify three forms of hybrid organizing: integration (multiple identities are shared by all stakeholders), differentiation (keeping the various groups and their views and identities separate), and a combination of both (Batillana et al., 2017, p. 139). Clearly, the development toward hybridization has merits and demerits. Organizational hybridity may lead to internal conflict, mission drift and challenges with respect to the organization's external legitimacy. On the positive side, organizational hybridity may bring a broader resource base, creativity and innovation power. Irrespective of the ultimate balance, grappling with organizational hybridity is one of the most pressing challenges the military, like other organizations, faces today. This becomes apparent in the following course of events.

The military and IT: The origins and unfolding of a case study[1]

After the attacks on 9/11, all NATO member states participated in the ISAF mission in Afghanistan, initiated by the United States. This was the consequence of Article V, the key to the treaty organization – "one for all, all for one". This participation implied deployments of large numbers of troops to warlike operations in unknown parts of the world, far away from home. For many nations' forces, this was a litmus test that required enormous efforts and resources. The Netherlands deployed a task force to the province of Uruzgan in the unruly south of Afghanistan, where it took the lead role in the operations in the area from 2006 to 2010 (Beeres, van der Meulen, Soeters & Vogelaar, 2012). This was an unprecedented challenge, an operation on a scale that the Netherlands military had not seen since the so-called "policing actions" in Indonesia, from 1946 to 1949. Not even participation in the UNPROFOR mission in the former Yugoslavia in the 1990s came close to this new experience, that is, in terms of the size of the workforce, the distance from home, and enemy threats.

The Dutch task force consisted of about 2,000 personnel, predominantly military personnel, and a number of civilians, among them diplomats from the Foreign Affairs Department. The task force contained infantry troops assembled in a battle-group, an air force component (Apaches, Chinooks, F-16s and fixed-wing transport aircraft), large logistics units and a so-called provincial reconstruction team. In the latter element, military officers played an important role too, but their efforts were enhanced by the diplomats and reserve officers. These officers were experts in different civilian domains, such as agriculture, education and architecture; they were tasked to help rebuild and further develop the Afghan infrastructure and economy in the region.

All reserve officers were deployed for a period of 6 months, after having been "militarized" for a number of weeks at the military academy and one of the garrisons at home. In some cases, specialized mission-oriented training and education were provided. After their deployment, the reserve officers resumed their professional activities and went back to working for their civilian employers. The practice of hiring specialists for temporary deployments was enhanced by the MoD through the development of an "employed support program" with a number of civilian employers, mostly firms in the business sector.

In one particular category of such reserve, officers were specialists who assisted the military with technical work, in particular information technology (IT), in the area of operations. Those reserve officers were urgently needed because the military did not have enough personnel in its workforce with such specialized IT competencies. The military had been experiencing the lack of such personnel as a pressing problem because operations are increasingly information-led. Hence, the solution of deploying reserve officers who were normally employed by for-profit companies in the IT field. This solution is a manifestation of the need for public–private partnerships in the provision and utilization of information technology in the public sector (e.g., Marschollek & Beck, 2012).

As explained earlier, all these reserve officers were "shared employees" as they all had two employers while being deployed: their civilian employer and the military. Civilian employers – as said, mostly for-profit companies – were interested in such sharing of the personnel because they wanted to support national efforts for ideological reasons, as a form of social entrepreneurship. At the same time, however, sharing personnel brought those commercial firms closer to the military. The military, in general, has ample means to assign large contracts, for which reason companies deem the military as offering an interesting opportunity to increase their turnover and profit. Reserves could in fact pierce the closed organizational boundary of the military organization. Once projects have been granted by the military, chances of success are likely to increase if companies have close ties and connections within the MoD. They simply know more and have developed interpersonal boundary-bridging relationships. For-profit companies, hence, have an indirect business interest in sharing their employees with the military; this outweighs possible downsides such as the loss of expertise during the reserves' absence and the risk their employees will return injured or not survive.

One of the reserve officers was tasked with assisting the military with monitoring IT challenges in Afghanistan. During his deployment, he drafted a report on what, in his eyes, needed to be done in order to improve the IT infrastructure of the Dutch task force in the area of operations. His report focused on the main camp and the various forward operating bases and particularly pointed at difficulties regarding information sharing with partners in the region.

Both his employers – the IT firm he worked for and the military – were satisfied with his analysis. Upon his return, the reserve officer was asked to present his findings at the MoD, in the presence of representatives of his civilian employer, the company he worked for. The email to which the presentation was attached had the words "business opportunities" in its heading. Despite this heading, the reserve officer was convinced his recommendations were important for the effectiveness and safety of the military operations. The email thus carried an ambiguous meaning.

The reserve officer continued working as a "shared employee" at the MoD, that is, as an advisor on a part-time basis, for some time. He spent the other part of his working time directly at his civilian employer's office. Everybody was more than content with his work. However, unexpectedly, a program was broadcast on national television regarding possible unethical practices by the MoD in tendering procedures. In particular, the IT firm that employed that particular reserve officer was mentioned. In this TV broadcast, the email drafted by the reserve officer and presented as "business opportunities" was shown. Apparently, someone from the IT company had, for personal reasons, leaked the email message to the broadcasting station on a USB stick.

The sequel was immediate and intense. The television program attracted nationwide attention and became the subject of a debate in the national parliament. This debate led to an in-depth study initiated by the Minister of Defense into the way the MoD acted vis-à-vis IT companies (Van der Steenhoven & Aalbersberg, 2016).

In the following days, the company that was accused of fraud lost almost 30 percent of its share value. The reserve officer who had performed to the satisfaction of all supervisors was suspended by the MoD, and his case was sent to the public prosecutor because of the suspicion that confidential information had been leaked. Additionally, legal specialists were tasked to find reasons to fire the reserve officer. The MoD argued that the company was responsible for the whole situation. In response to all the turmoil, the firm started its own investigations and it stopped supporting the practice of sharing reserves with the MoD. The company's management was not pleased with how the MoD had acted vis-à-vis their employee and the company. They threatened to go to court in support of their employee. Clearly, public and business logics fiercely clashed.

The investigation by the police and the public prosecutor was lengthy and did not result in a case, let alone a verdict; there was no evidence of any wrongdoing. In talks between the MoD, the employer and the reserve officer, it was decided that the reserve officer would be honorably discharged from his military duties. The firm recovered from the initial losses in share value, but its project turnover with central government agencies had dropped considerably. As much time had passed, parliamentary attention dropped, and the request to be informed about the end of the story was not repeated. In the meantime, the MoD tried to reinitiate its shared-employee policy with other firms. This had become necessary because other companies in the field had become hesitant about maintaining close cooperation with MoD, having seen what had happened to the company, a competitor, and its employee. The need for the military to tap expertise and knowledge from the business sector, however, remained as indispensable as ever.

Analysis: The strive for reputation control and the clash of logics

What happened immediately after the broadcasting of the allegations reflects what often ensues in times of (perceived) crisis. Boin, 't Hart, Stern and Sundelius (2017) studied the politics of crisis management and the role of public leadership under pressure. They distinguished a number of elements involved in managing a crisis, one of the most important elements referring to how strategic leadership makes sense of the crisis while it unfolds. What the organization often does reflects what the upper echelon thinks and does (Hambrick & Mason, 1984). Under the pressure of (perceived) crisis, political and strategic leadership oftentimes experiences the consequences of specific stress effects and cognitive limitations. The main effects are that decision makers focus on the short term, to the neglect of longer-term considerations. They fall back and rigidly cling to old and deeply rooted behavioral patterns; they rely on stereotypical behavior and are more easily irritable (Boin et al., 2017, p. 36). In general, under threat, there is a restriction of information processing (Staw, Sandelands & Dutton, 1981).

Looking back with hindsight, the impression arises that such reactions occurred at the apex of the MoD, following the television broadcast.[2] The reaction to submit the "case" to the public prosecutor had not been necessary from a legal point of view. Blaming the reserve officer's employer, with the consequential destruction of the cooperation with private-sector companies, was disproportionate in comparison with the allegations in the television program. There seemed to be a reason for such a reaction, however, and it had to do with the political vulnerability of the MoD's minister at the time.

Parliament had concluded in earlier cases that the central government in general was not in control of IT-related projects, as the problems also occurred in other ministries. The memory of previous fraud practices in tendering procedures, particularly in the construction industry, was still fresh (Sminia, 2011). Worries about similar problems also pertained to the MoD, and the allegations in the television program added to these concerns in parliament and society at large. In this context, the minister wanted to get this problem off her plate because the MoD's reputation was at risk – it is important to remember that (concern about) reputation offers a powerful explanation for a range of behaviors that have often escaped notice (Carpenter & Krause, 2012, p. 31). Hence, the blaming of the reserve officer and his employer. Given the specific and relatively weak position reserves (and contractors!) have, as we saw earlier, this was an easy thing to do.

The question, however, is to what degree the allegations may have been justified.

It must be admitted that the employer of the reserve officer had – perhaps unobtrusively and unintentionally – obtained a preferential information position vis-à-vis IT-related MoD projects in the near future. This is a violation of

public-tendering, logic-embracing equality. Despite the MoD's tendering procedures being conducted precisely in accordance with European law, informal practices may have been different. It was no secret that CEOs of for-profit companies
and high-ranking civil servants and generals regularly met at dinners organized
by the MoD to strengthen interorganizational relations. At these gatherings, the
issue a future *shared-employee covenant* was discussed, in a cozy atmosphere,
among friends from different societal sectors. All of this makes sense, but not to
everybody.

These dynamics relate to an observation made by Sarah Chayes (2006), an
American citizen living in Kandahar. She wrote about the way the US military
in Afghanistan and in Kandahar, in particular, at the time contracted host-nation
partners to assist them in multiple tasks. Those tasks pertained to transport, construction, security, logistics in general and linguistic services. Following the logics of economic organizing perfectly, these jobs were outsourced. Over time,
however, the US military favored only one regional political leader and his tribe
to do business with, which was, in Chayes's eyes, the Americans' major mistake
of the operation. The result was dramatic: there was no longer a level playing field
in the region because one partner was given preference all the time, at the expense
of other potential partners (tribes, political factions). This led Chayes (2006, p.
182) to conclude that, "it seemed to most Kandaharis that the primary mission of
US troops in Kandahar was to service Gul Agha Shirzai (the regional leader, with
whom the American partnered) and his Barakzai tribe". As a consequence, a feeling of resentment against the US troops developed and grew over time. Besides,
all relevant information was controlled by this one and only partner, excluding
others who became increasingly hostile. The mission's general legitimacy in the
region declined as a result (see also Soeters, 2018, pp. 53–4).

This observation demonstrates the US military's tendency to appreciate the
comfort and ease of having strong ties with a limited number of partners (e.g.,
Soeters, 2018). It also displays the weaknesses, the negative consequences, of a
strategy of preferred partnering. The best-known negative consequence, of course,
is the disproportionate, "strangling" pricing of services and goods by the supplier
who feels in a strong position (Porter, 1980, pp. 27–8). This has also occurred
in the military in action – for instance, in the US operations in Iraq, where a
contractor's operations caused excessive governmental costs, implying that the
governmental resources were used inappropriately (Hedgpeth, 2007). Chayes's
observations add to these downsides, as they indicate that having preferred partners, at the end of the day, may even lead to casualties among one's own troops.
Preferred partnering may result in lethal consequences.

The resemblance with our MoD case, not in operations but certainly related to
operations, is striking. In a study by Bidwell (2012), also on the outsourcing of IT
projects, it was shown that outsourcing decisions are often made in a much less
rational way than economic theory would predict (e.g., Soeters, 2020). Bidwell
demonstrated that senior and staff managers in a bank heavily promoted offshore
outsourcing and set targets to achieve this goal; at the same time, the project managers who were responsible for the daily results usually preferred to keep work

in-house, because that gave them better options to control the processes. This study showed that decisions about partnering and outsourcing, at least to a certain degree, are the result of internal politics and power dynamics between managers with conflicting preferences (Oliveira & Lumineau, 2019). There is much less rationality in such decisions than classical economic theory would predict.

In line with this finding, our case study provides an indication of power politics at the MoD concerning preferential partnering with civilian, mostly for-profit, partners. This time it is less about whether or not to outsource, but about which partner to choose. The case study indicates that, through the reserve officer's position, one firm gradually and informally may have obtained a preferential position, or at least may have made this impression. Casual remarks by MoD's top brass in talks with specific companies may create a false impression that ricochets in later formal tender procedures (Van der Steenhoven & Aalbersberg, 2016, p. ii). The study, besides, links sourcing challenges in peacetime and wartime conditions, and it shows that the perception of, and the response to, the unfolding crisis made the turmoil more damaging than was necessary. This "perfect" storm ruined the confluence of private and public capabilities in the field of IT-related defense, which was the whole idea behind the shared-employees program that the military organization needed so much for its operations at home and overseas.

In lieu of conclusions: Embracing paradox and possible repair of damage done

This case study is about how difficult it is to manage organizational hybridity. It is challenging for military organizations to deal with "strangers" and, at the same time, profit from the flexibility and diversity these bring into the organization. Via reserve forces, military organizations tap new resources, in particular, specific skills, knowledge, creativity and innovation power from the business sector. Yet, via reserve forces, private-sector logics enter the armed forces, which may cause unintended consequences such as a suddenly boiling crisis, as our case study showed.

Although, in this case, the IT company was suspected of improperly obtaining a competitive advantage, this turned out not to be the case. To prevent such events from happening again, it is crucial to carefully ensure that there is a level playing field for those who want to do business with the military. The relationship between formal tendering and informal relationships warrants more attention, as it seems messy and opaque. Maybe organizations need to embrace the paradoxes that surfaced in our case. Companies will always strive for differentiation and lock-in because of business continuity. Can the public organization resist, can it combine – as a hybrid – relational exchange with arm's-length tendering and contracting? It may have to if it engages in complex exchanges (see Ghoshal & Moran, 1996).

If there are no equal opportunities to compete for military projects and budgets, one way or another, problems are likely to emerge, be they legal or lethal. Experiencing fewer opportunities is likely to worry about organizations and the people therein, which may result in various disastrous outcomes, as we saw before. If the military wants to proceed with the *shared-employee* concept, it will need to continue to give access to all possible competitors, which implies, in this particular case, that reserve officers from all possible firms will be enabled to participate and possibly rotate. Even if this is already formally the case, one must make sure that this becomes an everyday reality.

Network dinners with CEOs from certain firms need to be replaced by network gatherings where all companies can participate. It is crucial to foster ties with many, not with a few. In such gatherings, uncomfortable realities (such as "only one or a combination of firms can win a tender" or "sharing personnel does not give any guarantee of doing business with us") should be presented as something to live with, not as something to avoid discussing (Smith & Besharov, 2019, p. 27). Most of all, no promises should be made to satisfy certain partners. It should also be made clear that cooperation between the MoD and certain companies is not fixed, but is flexible and may change over time, depending on the projects to come (Smith & Besharov, 2019, p. 29). Openness is the name of the game.

To repair the damage done vis-à-vis that particular company and, in fact, with the whole IT sector, the Netherlands military organization may want to engage in actions that enable dialoguing and co-creating (Petriglieri, 2015). Repair of relations can only be achieved through efforts from each side, from the side of the violator and the violated – in our case, respectively, the MoD and the company. The repair to damage in intra- and interorganizational relations involves responding positively to each other's repair attempts, which may be fostered by positive information disseminated by credible outsiders (Petriglieri, 2015). This, once again, shows that a good reputation will help to heal the pain of breaking up so abruptly and restore the opportunities provided by private–public cooperation in military organizations.

Notes

1 Specific data material on this case and the timeline of events can be obtained from the second author.
2 In another crisis in the same governing period, the political and strategic leadership at the MoD managed to find a temporary way out by denying any collateral damage and civilian casualties as a result of an NL-F16 bombing in Iraq, although reports about this mission's dramatic impact had already been published by Reuters. This denial, however, made the crisis go away only for a short period of time; it ricocheted years later when it became the subject of intense debate in parliament (see Bijleveld-Schouten, 2019). Clearly, there had been a focus on the short term, without considering the longer-term impact.

References

Bacharach, S.B., Bamberger, P., & Sonnenstuhl, W.J. (1996). The organizational transformation process: The micropolitics of dissonance reduction and the alignment of logics of action. *Administrative Science Quarterly*, *41*(3), 477–506.

Barley, S.R., & Kunda, G. (2001). *Gurus, hired guns, and warm bodies: Itinerant experts in a knowledge economy*. Princeton University Press.

Barley, S.R., & Kunda, G. (2006). Contracting: A new form of professional practice. *Academy of Management Perspectives, 20*(1), 45–66.

Batillana, J., Besharov, M., & Mitzinnick, B. (2017). On hybrids and hybrid organizing: A review and roadmap for future research. In R. Greenwood, Chr. Oliver, Th.B. Lawrence & R.E. Meyer (Eds.), *The SAGE handbook of organizational institutionalism* (pp. 128–62). Sage.

Beeres, R., van der Meulen, J., Soeters, J., & Vogelaar, A. (Eds.). (2012). *Mission Uruzgan: Collaborating in multiple coalitions in Afghanistan.* Amsterdam University Press.

Bidwell, M. (2012). Politics and firm boundaries: How organizational structure, group interests, and resources affect outsourcing. *Organization Science, 23*(6), 1622–1642.

Bijleveld-Schouten, A. (2019). *Beantwoording nadere vragen over de wapeninzet in Hawija* [Responses to additional questions about the use of force in Hawija]. Ministry of Defense.

Boin, A., 't Hart, P., Stern, E., & Sundelius, B. (2017). *The politics of crisis management: Public leadership under pressure.* Cambridge University Press.

Carpenter, D.P., & Krause, G.A. (2012). Reputation and public administration. *Public Administration Review, 72*(1), 26–32.

Chayes, S. (2006). *The punishment of virtue: Inside Afghanistan after the Taliban.* Penguin.

Cusumano, E., & Kinsey, C. (2015). Bureaucratic interests and the outsourcing of security: The privatization of diplomatic protection in the United States and the United Kingdom. *Armed Forces and Society,* 591–615.

Denis, J.-L., Ferlie, E., & van Gestel, N. (2015). Understanding hybridity in public organizations. *Public Administration, 93*(2), 273–289.

Farrell, T. (2010). Improving in war: Military adaptation and the British in Helmand Province, Afghanistan, 2006–2009. *Journal of Strategic Studies, 33*(4), 567–594.

Friedman, G. (2014). Workers without employers: Shadow corporations and the rise of the gig economy. *Review of Keynesian Economics, 2*(2), 171–188.

Ghoshal, S., & Moran, P. (1996). Bad for practice: A critique of the transaction cost theory. *Academy of Management Review, 21*(1), 13–47.

Hambrick, D.C., & Mason, P.A. (1984). Upper echelons: The organization as a reflection of its top managers. *Academy of Management Review, 9*(2), 193–206.

Hedgpeth, D. (2007, June 24). Iraq: Audit of KBR Iraq contract faults records for fuel, food: U.S. says it will increase monitoring in Baghdad. *Washington Post.*

Heinecken, L. (2014). Outsourcing public security: The unforeseen consequences for the military profession. *Armed Forces and Society, 40*(4), 625–646.

Jacobs, J. (1992). *Systems of survival: A dialogue on the moral foundations of commerce and politics.* Vintage Books.

Karpik, L. (1972). Les politiques et les logiques d'action de la grande entreprise industrielle. *Sociologie du Travail, 1*(1), 82–105.

Kelty, R., & Bierman, A. (2013). Ambivalence on the front lines: Perceptions of contractors in Iraq and Afghanistan. *Armed Forces and Society, 39*(1), 5–27.

Kirke, C. (2008). Issues in integrating territorial army soldiers into regular British units for operations: A regular view. *Defense and Security Analysis, 24*(2), 181–195.

Lomsky-Feder, E., Gazit, N., & Ben-Ari, E. (2008). Reserve soldiers as transmigrants. Moving between the civilian and military worlds. *Armed Forces and Society, 34*(4), 593–614.

Marschollek, O., & Beck, R. (2012). Alignment of divergent organizational cultures in IT public–private partnerships. *Business and Information Systems Engineering, 3*(3), 153–161.

Meyerson, D., Weick, K.E., & Kramer, R.M. (1996). Swift trust and temporary groups. In R.M. Kramer & T.R. Tyler (Eds.), *Trust in organizations: Frontiers of theory and research* (pp. 166–195). Sage.

Norheim-Martinsen, P.M. (2016). New sources of military change – Armed forces as normal organizations. *Defence Studies, 16*(3), 312–326.

Oliveira, N., & Lumineau, F. (2019). The dark side of interorganizational relationships: An integrative review and research agenda. *Journal of Management, 45*(1), 231–261.

Pache, A.-Cl., & Santos, F. (2010). When worlds collide: The internal dynamics of organizational responses to conflicting institutional demands. *Academy of Management Review, 35*(3), 455–476.

Pache, A.-Cl., & Santos, F. (2013). Inside the hybrid organization: Selective coupling as a response to competing institutional logics. *Academy of Management Journal, 56*(4), 972–1001.

Petriglieri, J.L. (2015). Co-creating relationship repair: Pathways to reconstructing destabilized organizational identification. *Administrative Science Quarterly, 60*(3), 518–557.

Pijpers, P.B.M.J. (2013). Outsourcing van militaire logistiek. *Militaire Spectator, 182*(6), 301–311.

Porter, M.E. (1980). *Competitive strategy: Techniques for analysing industries and competitors.* Free Press.

Sminia, H. (2011). Institutional continuity and the Dutch construction industry fiddle. *Organization Studies, 32*(11), 1559–1585.

Smith, W., & Besharov, M.L. (2019). Bowing for dual gods: How structured flexibility sustains organizational hybridity. *Administrative Science Quarterly, 64*(1), 1–44.

Soeters, J. (2018). *Sociology and military studies: Classical and current foundations.* Routledge.

Soeters, J. (2020). *Management and military studies: Classical and current foundations.* Routledge.

Staw, B.M., Sandelands, L.E., & Dutton, J.E. (1981). Threat-rigidity effects in organizational behavior: A multilevel analysis. *Administrative Science Quarterly, 26*(December), 501–524.

Van der Steenhoven, K.M., & Aalbersberg (2016). *Moral fitness. Een onderzoek naar de integriteit van de defensieorganisatie ten aanzien van de omgang met commerciële patijen op het gebied van IT.* The Hague: ABDTOPConsult.

Van Fenema, P.C., & Beeres, R. (2010). (Re-)drawing the boundaries: Sourcing operational and supportive services in military organizations. In J. Soeters, P.C. van Fenema & R. Beeres (Eds.), *Managing military organizations: Theory and practice* (pp. 84–96). Routledge.

Part V
Bringing it all Together

10 Integrative epilogue

What's new about the mission formation approach? Thinking through the military–academic juncture

Thomas Crosbie

As we have seen in Introduction by the editors, this volume is concerned with looking anew at contemporary military action or, in other words, the work of armed forces in theater. This epilogue argues for the vitality of this perspective by making fairly simple observations regarding both what is lacking in our existing knowledge and what is present and *new* in these collected chapters. The bottom line, which we might as well skip to now, concerns the special alchemy of the two words "mission" and "formation". An interrogation of missions, rather than functions or ecologies, strikes me as enormously productive and is most compelling when coupled with an analytically open approach to "formations".

Owing to the intense social forces acting upon deployed units, and the complex ways in which such units aggregate under hierarchical command to achieve delimited goals, mission formations have a special claim on our interest as social scientists. They are not quite the sum of their parts, never simply reducible to the will of the commanding officer or the policies governing (but only imperfectly restraining) them. Miguel A. Centeno and Elaine Enriquez, in their major synoptic work *War and Society*, note, "battle formations have some forms of emergent properties … produc[ing] a power greater than could be achieved through simple summation" (Centeno & Enriquez, 2016, p. 174). Nevertheless, military sociologists have a hard time letting go of their focus on the organizational forms, hierarchies and policies that look so very stable on paper. The mission-formations pivot offers a promising alternative to the standard thinking, capturing the rich sociological complexity of these entities as they evolve, fragment, reconfigure. As the chapters demonstrate, this shift allows us to acknowledge global phenomena – privatization, securitization, mediatization and so forth – without losing sight of the persistence of certain aspects of the military lifeworld and its professional and organizational logics.

Unfortunately, a new approach is badly needed. Consider the following historical irony. In the wake of disastrous unconventional operations in Southeast Asia, the US military pivoted to the fields of Europe with a renewed sense of purpose based on countering the technological advances made by Soviet forces in conventional operations. Forty years later, the US military is once again pivoting away from its failed unconventional operations to refocus on a conventional, Russian threat to its European allies. The first pivot was followed by organizational amnesia, as the US Army forgot how to counter insurgencies, ultimately contributing to a disastrous approach to wars in Afghanistan and Iraq. Will the second pivot be as costly? In other words, will we make the same mistakes a third time that we made in the 1960s and 2000s?

These questions demand that we interrogate the dark side of military innovation, the processes of organizational amnesia and suppression that plagued many militaries in the mid-century and may once again undermine the readiness of NATO militaries in the near future. But, of course, this is not the usual business of social scientists. As our disciplines responded to the cultural shifts of the 1960s and 1970s, a frequent casualty was the academic–military relationship. For some, this may have achieved the desirable end of protecting academia from the moral contagion of the military. Whatever we think of this, we must equally acknowledge that it came at the cost of mutual understanding.

This matters to us all. The first two decades of the twenty-first century have been marked by limited but still quite deadly forms of violent conflict, contributing to devastating refugee crises and, in ways we still do not fully understand, to the degrading of democratic structures at multiple levels. From my odd vantage point as an American-trained Canadian sociologist teaching and researching operations at a military college in Denmark, there is plenty of work to be done in bringing the profession into productive dialogue with the scholarship. Most troublingly, the new dynamics of war have largely outpaced NATO's analytic capacities, resulting in a new policy that fails to resolve the most pernicious problems that confront us and often leans on buzzwords and lazy abstractions rather than rigorous research (see, for example, Stoker & Whiteside, 2020).

Military organizations can hardly be faulted for failing to produce rigorous knowledge, given the paucity of support they receive from academia. Indeed, many scholars still view with outright suspicion any effort to help our own militaries better meet their challenges. Studying militaries is not always a prudent career move. Regardless, we must find ways to make this relationship work. By grounding us in the realities of military life as it is experienced in combat, the mission-formation approach promises to offer a realistic, accessible and immensely fascinating way into scholarship (for practitioners) and practice (for scholars).

During this long winter in the academic–military relationship, both scholars and military organizations have produced their "lessons learned" about modern conflict. What lessons did military leaders extract from these natural experiments in warfighting? More specifically, what lessons were learned (and integrated into doctrine) with respect to mission formations through the peacetime, wartime and technological innovations of the past two decades? Following this thread, we can start to unravel the real distance left to travel between scholarship about war and military preparations for future conflicts.

Academic lessons learned

The editors of this volume have provided a rough guide to academic lessons learned that can be compared broadly with military lessons learned.[1] Of course, militaries "learn" at multiple levels and in widely varying ways, but one type of learning has a special claim on our attention. If we agree with the editors that we want to better understand armed forces in theater, then we should focus a large share of our attention on the operational level of war, which is where decision-making authority tends to rest for most of the types of conflicts that interest us here. Further, to understand the operational level of war, we can focus on a sub-set of military policy referred to as "operational concepts", the most visible and immediately impactful elements of military policy.[2] Quite naturally, then, we arrive at a simple question to guide our understanding of whether academics and military thinkers are reaching similar conclusions: are the most robust academic lessons learned identified in this volume clearly expressed in the developmental tendencies of US operational concepts?

As a preliminary observation, and simply to answer this narrow question, we must bracket some of the rich insights expressed in the preceding pages, par-ticularly those that refer to tactical-level developments (e.g., the recognition by several contributors that contemporary war places an increasing burden on tac-tical commanders). Looking instead at mission formations at the larger end of the scale, we find frequent references among the contributors to seven distinct insights. First, we learn in this volume that large-scale missions are politically and socially complex. Second, they have fuzzy or fluid boundaries. Third, they have multiple logics of action. Fourth, they are emergent and provisional. Fifth, they take place within changing societal expectations. Sixth, they are amalgams of purely military and non-military elements. And, finally, they are shaped by new capacities opened up by new technologies.

Let us briefly tour through the empirical chapters of the volume to consider some of the proof already provided for these seven claims.[3] Rather than follow the order that the chapters appear in the volume, I reorder them here to tease out where each insight is most extensively analyzed.

Political and social

All of the chapters included here draw attention to the new political and social character of contemporary global conflict and reject many of our assumptions about the coherence and stability of military organizations as our preferred unit of analysis within these bewildering ecosystems. For Wilbur Scott, looking down from combatant commander General Stanley McChrystal's elevated position within this ecosystem, complexity is taken for granted. General McChrystal and staff were alerted to the complexity of the insurgency they were countering and utterly absorbed with the question of what to do about it. Jessica Glicken Turnley, writing about US military doctrine, focuses on an effort to think about complexity from a network analysis perspective. Eyal Ben-Ari, Joseph Soeters and his coau-thors, and Thomas Randrup Pedersen all focus on Northern European militaries deployed to Afghanistan, and, in each chapter, we see the political and social complexity of this operating environment shattering long-standing norms of pro-fessional military behavior. Uzi Ben-Shalom and Insoo Kim and Young-il Choi, writing about Israel and South Korea, respectively, equally stress the complex political questions that shape their countries' recent military experiences. Here, the authors are unanimous in their vision of contemporary militaries as acting in politically and socially challenging arenas.

Fuzzy and fluid

Jessica Glicken Turnley, discussing the US case, dwells upon the "fluid operating environment" in which military action occurs. Her case is network-centric warfare, which is explicitly an administrative response to overwhelming complexity. Fuzziness and fluidity pose a challenge to actors, but also to the researchers studying them. Uzi Ben-Shalom's chapter helpfully provides reflexivity on how we can make the complexity of contemporary war and conflict more tractable as researchers. Thinking through his own experiences studying the Israeli military, he reaches the conclusion that only multidisciplinary, methodologically diverse teams of researchers can hope to account for the complexity of our object of analysis.

Multiple logics of action

In his rich ethnography of Danish combat troops in Helmand, Thomas Randrup Pedersen reflects at length on military action as such. Taking the expression of violence as the presumed starting point, he highlights the competing logics – law, accountability, good order – that affect the tempo of a mission. Dipping down from the operational level to the experience of those individuals working at the tactical level, Pedersen provides an unsettling sketch of the frustrations and feelings of impotency that emerge when violent action is forestalled in the name of restraint.

Emergent and provisional ventures

Contingency, the state of being provisional, is a dominant theme throughout many of these chapters, but, in Eyal Ben-Ari's chapter, we are presented with a dizzying perspective on the missions undertaken by the military organizations. Taking as his starting point two volumes of documentation about Swedish military missions, Ben-Ari draws our attention to spontaneous instances of military organization manifesting throughout missions. Benefitting from the rich data collected by the Swedish Centre for Studies of Armed Forces and Society, he is able to question the scale at which mission formations occur, recommending in effect a fractal approach to military organization. This is a dramatic reconceptualization of traditional military sociology, dismissing the sacrality of the small unit which has occupied so much research attention in the history of the subfield and focusing instead on ad hoc and provisional configurations. The notion that mission formations are emergent and provisional ventures is a dominant theme in Scott's chapter about McChrystal as well. Famously, McChrystal was himself struck by the need to make a team of teams, his flexible organizational solution to the simultaneous need for specialists and generalized awareness.

Changing societal expectations

One productive avenue for reframing the significance of all of this complexity in our research topic is to use this to benchmark changing expectations within societies. Insoo Kim and Young-il Choi's chapter is a compelling example of how the taken-for-granted behaviors in this field can be unmasked as exceptionally odd developments. We have grown accustomed to the military-operations-other-than-war vision of military affairs, but consider how strange it is from a historical perspective that South Korean forces should be deployed on peacekeeping operations all over the world. This is surely a territorial army if there ever was one. Why is even this country optimizing its forces to be leaders in expeditionary warfare? The answer in this particular case is, in an important respect, not instrumentally rational, but rather densely cultural, embedding our understanding of operational behaviors in local norms and values. The theme of changing societal expectations resonates with Pedersen's chapter as well. The evolving cultural understanding of "warrior" is at the heart of his analysis. The young Danish men (mostly) who set out from their peaceful country to contribute to the International Security Assistance Force's open-ended mission in the Helmand province of Afghanistan were equally questing inward, asking themselves existential questions about their own identities. Although the identity of warrior may suffice sometimes for some of these individuals, more often it is just another "trap" that creates anxiety and frustration. Even hardened soldiers returning from combat find the old-fashioned identity of warrior hard to accept.

Military and hybrid amalgams

As Joseph Soeters, Gerold de Gooijer, Paul C. van Fenema and Nuno Oliveira show, organizational hybridity is easier said than done. Taking the Dutch military in the Uruzgan province of Afghanistan as their case, they explore the unexpected consequences of employing reservists as "shared employees", a dual-hatted role allowing them to serve simultaneously as uniformed officers and as civilian employees of an IT company. One such employee brought a corporate logic to his assessment of the battlefield in a report to the Ministry of Defense. Once leaked, his report scandalized the Dutch public, who began to question the profit motive driving military interventions. One take-away is that military–civilian hybrid organizations have fault lines undermining their stability. Equally, the authors reveal that the new organizational forms taken by militaries are not unique to militaries and may better be understood as a decline in military uniqueness. We see the transformation of military forces into more agile, mission-focused organizations, but this is also happening throughout the public sector through the new public management reform wave. Just because managers want militaries to conform to civilian norms does not mean they will. Here, culture again trumps instrumental rationality.

New technology

As with the first insight, this final insight is shared unanimously by the contributors. Jessica Glicken Turnley provides perhaps the most sustained discussion, exploring the bizarre pathways by which the US military has attempted to integrate new insights into networks into warfighting doctrine and organization. There is nothing straightforward or simple about integrating new technologies into military force projection. Through complex bureaucratic processes, technologies do finally make their way through the organization and have an impact – often, not the impact we might want or expect.

The preceding chapters have treated these insights in much greater depth and with far more nuance than is possible here. My goal is not to repeat what has already been said (and briefly summarized at the end of Introduction), but rather to gather some common themes for the purpose of comparison with military lessons learned. For this reason, I will slightly rephrase these seven academic insights into categories of interest to military policymakers.

First, conflicts are politically and socially challenging, gesturing toward the increasing importance of *foreign politics*. Second, they have fuzzy or fluid boundaries: *complexity* is a comparable major theme in military thinking. Third, they have multiple logics of action: in military terms, these missions are not *autonomous* (controlled by military concerns alone). Fourth, they are emergent and provisional, which again is discussed in terms of autonomy. Fifth, they take place within changing societal expectations, which we can view as a concern with *domestic politics*. Sixth, they are amalgams of purely military and non-military elements: again, in military terms, this matches with autonomy concerns. And, finally, they are shaped by new capacities opened up by new *technologies*, which is also the language used in military writing.

Simplified in this way, a productive new line of inquiry emerges from the mission-formations perspective. If we follow the mission-formations approach, we arrive at seven scholarly insights that can drive future research and analysis. However, we can also compare the scholarly vision with the lessons learned by militaries. In doing so, we can simplify our search to only four major categories of concern which we should expect to find in military documents: first, a focus on the importance of political concerns (both foreign and domestic); second, a recognition of the challenge of increasing complexity; third, preparation for the declining autonomy of military action; and fourth, active embracing of new technologies. Table 10.1 sketches this shift and highlights a simplified way of thinking about academic lessons learned in terms immediately recognizable to military policy readers, with repeated items crossed out.

Are military policymakers coming to the same conclusions as scholars? If so, then we should expect new military policy will grapple in earnest with the importance of politics, the increasing complexity of military affairs, the military's declining autonomy from other elements of national power, and the need to appropriately integrate new technologies. This modest list is one eminently practical outcome of the mission-formations project.

Table 10.1 Scholarly Lessons Learned as Equivalent Military Policy Concepts

Scholarly Language	Military Policy Equivalent
political and social	(foreign) politics (1)
fuzzy and fluid	complexity (2)
multiple logics of action	loss of autonomy (3)
emergent and provisional ventures	~~loss of autonomy~~ (3)
changing societal expectations	(domestic) politics (1)
military and hybrid amalgams	~~loss of autonomy~~ (3)
new technology	new technology (4)

Let us look very briefly at the newest operational concept to gain high-level investment. It is in the nature of operational concepts that they remain largely obscure to researchers until they are finally declassified. Nevertheless, a new concept has emerged within the US Department of Defense that is unusually public-facing. This concept began life within the US Army's Training and Doctrine Command under the title Multi-Domain Battle. It gradually evolved into Multidomain Operations (MDO) and has very recently been renamed All-Domain Operations (ADO). While ADO remains inchoate and classified, MDO's impact was felt in the 2017 revised edition of the US Army's Field Manual 3-0, *Operations* (FM 3-0).

Do we see the four critical insights about war reflected in this manual? The answer is overwhelmingly yes. Throughout the manual, the authors stress that war and peace are a continuum that future operating environments will be complex and chaotic, that soft power is absolutely essential to future operations.

I suspect that any casual reader will be struck by how closely new military doctrine aligns in principle with the cutting-edge insights raised in this volume. However, differences between military and academic insights are no less obvious, and it is here where a mission-formations perspective can be most helpful. Despite the provisional nature of missions, they stand atop enormous, rigid bureaucratic structures which supply and sustain them in combat, and which train and deploy them to combat. These are also the structures that produce the policies that inform behavior, the doctrine and concepts that are so determinative of courses of action. Thus, the central insight of the mission-formations approach (to disaggregate the organizations that exist in the mission from the organizations that exist outside the mission) is inevitably absent from manuals such as FM 3-0. Once deployed, the army (and equivalent organizations) still structures thinking and behavior but cannot structure reality itself. Complex missions will spin out of the doctrine's and the organization's grasp, and what is left is the messy reality that professional officers must negotiate in partnership with a wide variety of other, as yet unknown, participants. And thus, as much as operational concepts are flexed to acknowledge complex political and social realities, the lessons about the loss of military autonomy will be the hardest to learn – and the easiest to forget.

Conclusions

I began this short epilogue asking whether the pivot currently underway in US defense planning away from countering insurgencies and back to near-peer competition will be as costly as the same pivot done 40 years ago. Are we repeating our mistakes? I do not provide an answer, but nor has the field generated an answer. It seems to me that social scientists should occupy a great deal of their time with questions such as this. As those of us who work on military-adjacent topics know, however, questions as wide-ranging and consequential as this are virtually ignored in the literature. There are simply too few people working on military topics to cover all of the ways in which research should inform – and hold accountable – policy decisions pertaining to the projection of force. Instead, by focusing on high-level strategic concerns about *when* and *if* wars should be fought, the social sciences have largely ceded questions of *how* they should be fought to militaries. By turning our back on the minutiae of how wars are fought, we ignore many critical questions about the use of legitimate violence, which are thereby left to militaries to resolve on their own.

The mission-formations intervention is not simply a reworking of old academic ideas, a post-postmodern approach to the military. Indeed, military organizations are not necessarily the subject of this research at all. We are invited instead to study *missions* and whatever collections of human beings happen to form around those missions. In this sense, the editors of and contributors to this volume provide a provocative new approach to studying the violence done in our name that breaks the longest-standing norm in the field – to take for granted the administrative logics of the field we study. By thinking analytically about how missions link across multiple bureaucracies and occupational groups, the mission-formations approach invites scholars to describe a modern war in our own language. At its core, the mission-formations pivot is a turn to face war as it exists today.

Notes

1 I am grateful to Eyal Ben-Ari, Uzi Ben-Shalom, Thomas Brønd and Carmit Padan for their identification of these common academic insights, listed below.
2 "Operational concept" is a term of art broadly encompassing defense-sector policy configurations that articulate distinct approaches to projecting military force. Owing to the nature of the alliance, NATO countries tend to be passive adopters of US operational concepts (Bjerga & Haaland, 2010).
3 I leave aside here the theoretical and syncretic chapters at the beginning of this volume.

References

Bjerga, K.I., & Haaland, T.L. (2010). Development of military doctrine: The particular case of small states. *Journal of Strategic Studies, 33*(4), 505–533.
Centeno, M.A., & Enriquez, E. (2016). *War and society.* Chichester: Wiley.
Stoker, D., & Whiteside, C. (2020). Gray-zone conflict and hybrid war: Two failures of American strategic thinking. *Naval War College Review, 73*(1), 19–54.

Index

Abizaid, John (US General) 51
academy and academic 15, 25, 109–111,
    118, 137–138, Chap 10
actionable knowledge 91
adaptation 13, 16–17, 29, 31, 35, 37, 39,
    45, 53–54, 58, 77, 98, 109, 154
Afghanistan 6, 16, 25–26, 33, 39, 45–46,
    49, 53, 55, 59, 73, 91, 93–94, 125, 127,
    132, 134–135, 139, 141–144, 146–147,
    155, 161–162, 167, 169, 171–172, 175,
    184–185, 187
Air Force 49, 58, 64, 72, 96, 115, 147, 171
Algeria 53–54
Armaments 31; *see also* weapons
anthropology 4, 16, 25, 33–34, 118,
    141–145, 154, 161–162
asymmetrical conflicts 47, 64, 141
artillery 5, 108, 112, 114
authority 12–13, 51, 55, 58, 67–71, 74–77,
    80–86, 98–100, 11, 115, 156, 170, 185

battle groups 5, 148–156, 171
battlespace 5, 25, 35, 55, 72, 144, 146,
    150, 161–162
Ben-Ari, E. 4–8, 9–12, 14–16, 47–48,
    52–53, 56–57, 61, 64, 93, 100, 102,
    108–109, 142–144, 151, 155–156, 163,
    170, 185
Ben-Shalom, U. 8, 14–16, 58, 92, 101, 107–
    111, 113–116, 142–143, 151, 163, 186
bonding *see* cohesion
Brønd, T. V. 142, 151, 163
bureaucracy and bureaucracies 13–14,
    Chap 4, 100, 107, 188–189

Canada 23, 25, 31, 33
careers and careering 45, 49, 169, 185
casualties 10, 30, 53, 154–155, 159,
    174, 185

casualty aversion 6
centralization 50, 81–85
Civil-Military Coordination (CIMIC),
    South Korea 131, 132, 148
chain-of-Command 55, 58, 94, 156, 170
China 24, 48
Choi, Y-Il. 15, 185, 187
cities 49
citizens 66, 144–145, 168, 175
civilian contractors 10, 16
civil-military relations 6, 29, 99
Civil war 46, 49, 123, 125
coercive isomorphism 133, 134, 136
cohesion 2, 7–9, 12–13, 55–58, 99–170
COIN 69, 94, 144–145, 157, 166
collateral damage 10, 155, 160, 177
collective action 9, 13, 17, 100, 144, 146
combat 4–6
combat formations 10, 13–14, 24–34,
    37–39, Chap 5, Chap 6, Chap 7
combat units 6–9, 11, 13–14, 16–17, 64,
    Chap 5, 143–144, 161
combatants 61, 114, 133
command 5, 13–14, 32, 58, 67–68, 74–77,
    82, 87, 92–93, 96–97, 111–112, 117,
    148, 153, 155–156, 161–170, 183;
    see also leadership
command and control 38, 72, 83–87, 97, 149
communication 2, 5, 7, 13, 50, 56, 72, 75,
    78, 83, 85, 96, 98–100, 108–109, 112,
    128–129, 157
complexity and complexities 9, 11, 26, 36,
    38, 81–82, 85, 97, 108–109, 111, 185–189
conscripts 47, 108, 114
conventional wars 3, 10, 16, 39, 47–48, 61,
    123–124
cooperation 5, 12, 16, 24, 48, 56, 79, 94,
    96, 99, 115, 125, 131–132, 136–138,
    158, 173–174, 177

coordination 4–5, 9, 12, 14, 18, 81, 83–89, 102, 107, 111–113, 118, 152, 154
counterinsurgency (COIN) 25–26, 28, 59, 148–149, 158, 161, 164; *see also* insurgency
courage 10, 99; *see also* bravery
Crosbie, T. 17

death 16, 60, 110–111, 140–142, 153–154; *see also* killing
decentralization and decentralized 51, 58, 66, 73, 84, 100
decision and decision-making 7, 10, 13, 27–28, 30, 35, 47, 55, 58, 69, 76, 114, 128, 151, 175–176, 185, 190
democracy, Chap 8
Denmark 147, 149, 163, 184
discipline (military) 8–9, 11
division of labor 95, 99–100
doctrines, military 12, 36, 59, 72–76, 107–108, 112–113, 116, 118, 126, 142, 146, 152–156, 161–162, 164–165, 188–189
Durkheim, E. 26–27, 66, 69

effective and effectiveness 7–8, 16, 18, 28, 33, 35, 57–58, 061, 67, 81–82, 85, 87, 91, 109, 114, 144, 155, 157, 162, 167, 169, 173
Egypt 45, 114
Elias, Norbert 179
emergence 6, 8–9, 12, 14–15, 71, 86, 132, 142, 144, 161, 169, 183, 185–186, 188–189
emotions 32, 54, 56, 58, 69, 93, 141
employees 16
enemy and enemies 34
Esprit-de-Corps 58
ethical 4, 16, 53, 66, 143, 169, 173

family and families 56; *see also* parents
fieldwork 16, 118, 141–143, 145–146, 161–162; *see also* methodology
fluidity 73, 85, 107, 110, 186
friction of battle 8
functionalism 8, 12, 24, 33
fuzzy (conflicts) 98, 185–186, 188–189

gatekeepers 111
Gaza 109, 116–117
Gazit, N. 6, 118, 170
gendarmerie 96, 98
gender and gendered 32, 93, 111, 114–115

Germany 23, 33
global surveillance 10, 52–53
Global War on Terror 46–47, 125, 134; *see also* War on Terror
Gooijer, G-De. 16, 187
ground forces 14, 16, 46, 64, 108–109, 115–116, 143, 150

"hearts and minds" 112
heroes and heroism 135, 157
hierarchy and hierarchical structures 7–9, 12, 14, 16, 51, 64, 67–70, 72–75, 81–83, 85, 93–94, 100, 107, 153, 167, 183
humanitarian missions and assistance 3, 10, 24, 53, 61, 82, 88, 92, 96, 108, 125, 129–130, 137
human rights 6, 10, 53, 123, 125
Hussein, Saddam. 46, 49–50
hybrid conflicts and conflicts 91, 123
hybridity 5, 17, 98, 123, 144, 168–169, 171, 176, 187, 189

identity 179, 187
imagined community (B. Anderson) 58
implicit knowledge 34
improvisation 9, 91, 111; *see also* innovation
in-between organizations 53, 170
incident command system 14, 67, 80–86
Indochina 53–54
informal psychological contracts 14, 101
informal relations and communication 16, 31, 75, 81, 85, 100, 176
information technology (IT) 16, Chap 9
injuries 38, 172; *see also* casualties and wounds
innovation 4, 17, 28, 31, 35, 52, 75, 78, 81, 85, 170–171, 178, 184
instant units 57
institutional analysis 13, Chap 2, Chap 12
institutional environment 15, 132
institutional isomorphism 132, 133
institutional logics 27, 37–39, 129
insurgency 60–61, 78, 184, 186, 190; *see also* counterinsurgency
intelligence, military 5, 28, 49, 51, 54–5, 58, 73, 79–80, 151, 157, 163
intelligence agencies 54
interpreters 5, 93–94
Iraq 6, 13, 26, 33, 39, 45–56, 59–60, 78–79, 109, 125–127, 134–135, 137, 167, 169, 175, 184
irregular Warfare 3, 29

ISAF 16, 93–94, 141–143, 146–148,
    150–153, 155–156, 160–163; *see also*
    NATO
Israel 25–26, 29–30, 47–48, 53, 57–58,
    108–109, 113–114, 117–118, 186

Janowitz, M. 7
Japan 6, 45, 85
jointness 5, 92
joint task force (JFT) 14, 72
judiciary 10
juridification 6, 156, 161

killing 8, 49, 59, 110, 114, 160, 163;
    see also death
Kim, I. 15, 185, 187
King, A. 5–8, 11, 107, 142, 144–145
Korea, Republic of 123–125, 128,
    133–134, 139

laborers 16
law and lawyers 7, 10, 27, 59, 61, 67, 99,
    125, 175, 186
leadership and leaders 111, 113–114, 154,
    169, 174–175, 177; *see also* command
Lebanon 29, 45, 109, 127–133, 135–137
legitimation and legitimacy 6, 8–9,
    28–30, 72, 87, 100, 124, 132, 170–171,
    175, 190
lessons learned 39, 45, 184–185, 188–189
Levy, Y. 6, 10, 15, 108
Libya 32, 39
Little, R. W. 111
logics of action 6, 16–17, 96, 167–172,
    174–175, 186, 188–190; *see also*
    institutional logics
loose-coupling 11–12, 14, 51, 57, 84,
    Chap 5

market and marketization 6, 38, 52, 65,
    70, 72
Matthews, D. 141
Matthews, M. D. 3
McChrystal, S. (US General) 13, 45–46,
    49–60, 73, 79, 80, 84, 85, 87, 185–187
media and mediatization 10, 32, 53, 65, 93,
    97, 99, 100, 113, 114, 183
memorials and memorialization 146
memory 174
methodology 32, 34, 36, 39, 111, 118, 143,
    152; and participant observation 143
militarism and militarization 172
militarization 172
military casualties *see* casualties

military reserves 16, 168, 170, 173, 174
military service 11, 14, 72, 73, 109, 161
military socialization *see* socialization
military sociology, Chap 2., 4, 7–8, 12–19,
    23, 144, 186
military units, textbook units 92; battalions
    7, 92, 102, 144; companies 7, 10, 17, 92,
    96, 102, 144, 161, 118; platoons 6, 7,
    9, 69, 92–93, 109, 114, 126, 142, 144,
    150–161; squad 7, 9, 114, 115, 141, 143
mimetic isomorphism 133
minorities 24
mission command 58, 61, 154–155
modularity and modular units 91, 93
morale 12, 142
Moskos, C. C. 6–7, 111, 144
motivation 3, 7, 11, 14, 23, 52, 58, 91, 142,
    144, 162
multinational missions and forces 4,
    125, 129
Munkler, H. 3, 91
mation-state 48, 64–65, 70–71

NATO 2, 13, 151, 184; and doctrine
    153–156; *see also* ISAF, KFOR, etc.;
    and research projects 32–33
Naveh, S. 26
the Netherlands 16, 23, 33, 168, 171, 177
network-centric warfare 75, 186
networks Chap 4, 56
new wars 2, 4, 9, 52, 92, 123, 151, 154
non-governmental organizations (NGOs)
    10–11, 73, 75
nonstate actors 3
normative isomorphism 133, 136
North Korea 123–124, 130, 133
Norway 33

Obama, B. 45, 60
Oliveira, N. 16, 176, 187
organic mission formation 128, 137
"organic unit" 3, 5–6, 14, 17–18
organization science 4
organization studies 35, 37
Ouellet, E. 8, 12, 15, 18, 23, 26, 28–29
outsourcing 167, 169, 171, 175–176

Padan, C. 109, 142, 151, 163, 190
participant-observation *see* methodology
peace-keeping and peace-enforcement 123,
    125, 127
Pedersen, T. R. 16, 141–143, 145, 149,
    155, 162, 187–195
Petraeus, D. (US General). 59

planning Chap 3, 13, 51, 77, 80, 94, 155, 162, 190
police and policing 3, 5, 50, 60, 69–70, 96, 103, 106–109, 123, 127, 134, 140
politics 62, 174, 176, 188
"post-heroic warfare" 3
"postmodern militaries" 144
precision warfare 156
press *see* media
prestige 11, 29
Private military companies (PMC) 167
privatization 10, 178, 183
professionals and professionalism (military) 136, 169
provincial reconstruction teams (PRTs) 92, 95, 98, 125, 148, 158
psychological contract 101
psychology and psychologists 4, 8, 36, 39, 58, 117

recruitment and retention 12, 107
reflexivity and reflectivity 26–27, 34–36, 111, 117, 186
research design 108; *see also* methodology
reserves and reserve service *see* military reserves
resilience 7, 11, 17, 26, 146
"risk-transfer-war" 10
rites and rituals 107; *see also* ceremonies
Russia 23–25, 32, 48, 183

Scott, W. J. 13, 185–186
Segal, D.R. 57–58
sense making 109
Shamir, B. 92, 100
Shamir, E. 12, 58, 109
shared consciousness 54–56
Shields, P.M. 92, 110
small wars 48, 52, 58
socialization 4, 8, 17
social psychology 4
social science and social sciences 17, 30, 109, 190
sociology *see* military sociology
Soeters, J. 5, 8–9, 16, 99–100, 105, 107, 110, 118, 168, 171, 175, 186, 189
soldiering, "soldier-diplomats" and "soldier-warriors" 16

solidarity *see* cohesion
special operations forces (SOF) and special forces 14, 34, 64, 71
strategic corporal 11
stress 7, 174
suicide bombers 47, 49, 53
Sweden 33, 91, 94, 97, 100
swift trust 57, 101

task cohesion *see* cohesion
task forces 38, 45, 72, 92, 136, 144, 181
"Team of Teams" 15, 54
technology 31, 41, 65, 70, 75, 182, 188
television *see* media
terror and terrorism 46, 58, 87, 113, 125, 135, 147; *see also* War on Terror
"textbook units" 5, 144
TF 714 (Task Force 714) 45
training 11, 33, 46, 76, 94, 98, 107, 112, 117, 125
transnationalism and transnational ties 70
trust 54, 70, 84, 95, 101, 126, 170; *see also* swift trust
Turnley, J. T. 13–14, 64, 67, 78, 185–186, 188

United Arab Emirates 45, 125
United Kingdom 23, 33
United Nations 123, 125, 127
United Nations Interim Force in Lebanon (UNIFIL) 131, 136
United States 23, 33, 35, 64, 76–77, 87, 158, 167, 171

Van Fenema, P.C. 4, 8, 16, 99, 167, 187, 197
veterans 107
Vietnam War 133, 138
volunteers and volunteering 53, 84

"Warfighting ethos" 142
War on terror 46, 125, 134
weapons and weapon systems 8, 124, 150, 156
Weber, M. 67
Winslow, D. 10, 99, 110
World War Two 7, 38, 47, 56, 123